P9-DFN-989

CARVING GRAND CANYON

View west from Yaki Point. Photograph by George H. H. Huey

CARVING GRAND CANYON

Evidence, Theories, and Mystery

Wayne Ranney

Grand Canyon Association
Inspire. Educate. Protect.
Grand Canyon, Arizona

Copyright © 2005, 2012 by the Grand Canyon Association
Text copyright © 2005, 2012 by Wayne Ranney
All illustrations are the property of their respective artists and are protected by copyright law.

First edition published 2005. Second edition published 2012.

All Rights Reserved.
No portion of this book may be reproduced in whole or in part, by any means (with the exception of short quotes for the purpose of review), without permission of the publisher.

Cover: Sunrise from Mohave Point on the South Rim, with Colorado River in view.
Photograph by Mike Buchheit
Back cover: Butte Fault. Photograph by Wayne Ranney
Half title page: Sunset on the Colorado River near Nankoweap. Photograph by Larry Lindahl

Edited by Todd Berger, Faith Marcovecchio, and Andrea Rud
Designed by Jamison Spittler, Jamison Design
Maps and diagrams by Bronze Black
Indexed by Sandi Schroeder, Schroeder Indexing Services
Printed in China

Second Edition
16 15 14 2 3 4 5

Library of Congress Cataloging-in-Publication Data
Ranney, Wayne.
Carving Grand Canyon : evidence, theories, and mystery / Wayne Ranney. --
2nd ed.
p. cm.
Includes bibliographical references and index.
ISBN 978-1-934656-36-5 (alk. paper)
1. Geology--Arizona--Grand Canyon--Popular works. 2. Geology--Colorado River (Colo.-Mexico)--Popular works. 3. Grand Canyon (Ariz.)--Discovery and exploration--Popular works. I. Title.
QE86.G73R36 2012
557.91'32--dc23

2012015001

The mission of the Grand Canyon Association is to help preserve and protect Grand Canyon National Park by cultivating support through education and understanding of the park. Proceeds from the sale of this book will be used to support research and education at Grand Canyon National Park.

Grand Canyon Association
P.O. Box 399
Grand Canyon, AZ 86023-0399
(800) 858-2808
www.grandcanyon.org

THIS BOOK IS DEDICATED TO two institutions that I support very much—the Museum of Northern Arizona and the Grand Canyon Association. In 1976 they co-published the "Geologic Map of the Eastern Part of the Grand Canyon National Park, Arizona" (with revisions in 1986 and 1996). This colorful map has served to remind me every day that the Grand Canyon and all that it represents is there always to inspire and enchant. A unique blend of science and art, this map has been the primary motivator for my interest in Grand Canyon's origin, and without its publication, this book might never have seen the light of day. Both the Museum of Northern Arizona and the Grand Canyon Association, whose missions are to increase and share knowledge about the Grand Canyon region, deserve every ounce of support we can give them.

Detail from *Point Sublime* by William H. Holmes, 1882.

CONTENTS

FOREWORD

Site and Insight: The Deep History of an Idea

Even the most casual observer soon recognizes that the Grand Canyon is the outcome of an uplift and a river. The uplift is the series of plateaus—four in all, the Shivwits, the Uinkaret, the Kanab, the Kaibab—that rise like steps from the cliffs at Grand Wash to the Kaibab monocline. The river is the Colorado. Somehow river and uplift cut down and rose up through one another. Those gorges subsequently widened, with each plateau displaying a distinctive variant on that common theme. Collectively, the cascade of resulting canyons made the Grand Canyon.

The outcome has long been celebrated as a testimony to deep history. To understand the canyon requires an imagination that can see the earth's ancient past in its exposed rocks; to explain the canyon is to tell a story that reaches back to times before today's continents existed, that narrates the ascent and decline of geologic empires, each rising on the eroded ruins of its predecessor. The strata in the canyon's cliffs seemingly describe that text for all with eyes to read.

But deep history is an idea, and ideas, too, have their history. The rocks could not be viewed as a geologic spectacle until the science of geology was invented. You can't read without knowing the language and its written alphabet. The exposed rocks of the inner gorge may be nearly 2,000 million years old, but geology as a science capable of deciphering them is little more than two hundred. The earth sciences furnished the spectacles by which to view a supremely visual scene.

Even so, a space becomes a scene only when there is someone present to see it. The special value of *Carving Grand Canyon* is that it matches viewer with view. It not only lays out the geologic processes that are believed to account for the landscape but also describes those geologists who reconciled place with idea, who saw deep history in its gorge, ideas in its rocks, and meaning in the collision of culture with canyon. They made space into scene and scene into spectacle.

The geographic place that is Grand Canyon has been known to Western civilization since Capt. García López de Cárdenas of the Coronado expedition stood on the rim in 1540. The canyon, in fact, was the earliest of

North America's natural wonders to be discovered, and the Colorado River the first of its great rivers to be mapped. The canyon predates Hennepin's discovery of Niagara Falls by 148 years and Meek's encounter with Yosemite by 293. The Colorado was explored 147 years before LaSalle ventured down the Mississippi. It was rediscovered in the eighteenth century and again in the early nineteenth. But it had no obvious value. It housed no wealthy peoples ready for conquest or conversion, held no gold placers or beaver pelts, and could not serve as a corridor of travel (on the contrary, it was a barrier). It was, according to standard geopolitical calculus, worthless. That assessment was repeated by visitor after visitor through 1857–58, when the Army Corps of Topographical Engineers sponsored an expedition under Lt. Joseph Ives, who pronounced it "altogether valueless." It was simply the Big Cañon. The smart money said it would always remain of little worth.

Geology—an emerging discipline that only got its name in 1783—said otherwise. For anyone curious about earth history, the Grand Canyon was a "paradise." That was the verdict of John Strong Newberry, physician-naturalist with Ives. The chasm between the perspectives of Ives and Newberry was as wide as the distance between rims. Others visitors followed—geologists most prominently, but also landscape artists, litterateurs, cultural commentators, nationalists. In 1875 John Wesley Powell, leader of the first party through the canyon, published *The Exploration of the Colorado River of the West*. That expedition unveiled the last unknown river and mountain range in the United States and transfigured a "Big" canyon into a "Grand" one. In 1882 Clarence Dutton published the *Tertiary History of the Grand Cañon District* as the first monograph of the fledgling U.S. Geological Survey. Together, those two books gave the canyon its foundational views from river and rim. Twenty-one years later, President Teddy Roosevelt rode a spur line of the Santa Fe Railroad to the South Rim and declared the scene the "one great sight every American should see."

For much of the century that followed, the canyon was an exemplar of geology. It adorned textbook covers and served as a lithic new testament to the scientific understanding of creation. It was a place where a geologist could make a reputation, and few major figures (particularly Americans, as Wayne Ranney records) failed to contemplate rim and river.

Yet, curiously, an explanation for its own creation remained elusive. The place was incontestably about uplift, river, and erosion, but how they came together remained unsettled. Was the land first, or the river? Dutton even spoke of a Law of the Persistence of Rivers. The flowing river, it seemed, endured; the solid lands around it warped and eroded. So, too, the flow of those early ideas persisted while the landscape of ideas—whole syndromes of thought—bulged up and wasted away. Throughout, the prevailing concepts only entrenched more deeply.

Still, the defining feature—the informing event in Grand Canyon history—is the odd bend the Colorado River takes as it angles sharply westward and cuts against the grain of the plateaus. It was hard to explain why a river would do this, even if its meanders predated the upwelling that would strip off ten thousand feet of Mesozoic sediment and allow the river to incise its gorge. From time to time new theories emerged. Some graded into one another, like limestone segueing into shale. Others arose on unconformities in the record of scientific thinking.

The breakthrough arrived during a 1964 symposium chaired by Eddie McKee and sponsored by the Museum of Northern Arizona. The meeting came midway between the centennials of the Ives and Powell expeditions that had first subjected the canyon to scientific inquiry. The critical insight was that the Colorado River had, in fact, its own history. Canyon and Colorado had co-evolved. Headcutting, downcutting, overspilling—as landscapes flexed and twisted and wore away, such were the processes that allowed the fluvial pieces to come together into its modern configuration. The Grand Canyon was not the result of one fixed process acting on another but of two changing processes mutually interacting.

A similar process, and mystery, surrounds the revelation of the canyon as a cultural spectacle. The canyon was not intrinsically there, an enduring natural wonder awaiting discovery. It was a landscape that had to be sculpted into meaning, and the wonder is how a civilization that had for centuries deliberately avoided or dismissed the place suddenly veered through it in apparent defiance of inherited notions of landscape values, aesthetics, and geopolitical significance. The most likely explanation is that, like the Colorado River, the cultural mainstream was the composite of many tributaries integrated by the intellectual equivalent of headcutting, downcutting, overspilling, and capture.

The Grand Canyon, in brief, was not the outcome of an awe-inspiring landscape lying dormant until found and incorporated into American society, but the unexpected by-product of place and culture interacting in complicated ways. Its scientists made the canyon worth examining, its exploring artists made it gorgeous, and its institutionalization into a park led a curious public to make it into a national shrine. Moreover, all this happened with a suddenness—within a handful of years—that seemed to emulate in the culture the typical reaction of most visitors to the rim. The canyon was instantly there. It seems it must always have been so.

It is the particular value of Wayne Ranney's gracious and visually enthralling book that he traces the parallel histories of geologic gorge and perceived canyon. The physical canyon is the outcome of long erosion. How this might have happened is the subject of a careful exposition of geological processes and the odd way they can combine. The cultural canyon is the outcome of a long evolution of social encounters. He parses the scene and gives its oft-overlooked partisans recognition, much as a geologist might break down the canyon cliffs into distinctive strata, and then reassembles them into a coherent story. Today no one can look at the canyon without appreciating the deep time it represents. But neither can anyone read contemporary explanations of the canyon without appreciating the patient upheavals and erosions of ideas that have made that understanding possible.

Clarence Dutton famously declared that the canyon was grand not by virtue of any one of its features, no matter how fabulous, but by virtue of its whole, its *ensemble*, the way it brings together gorge, cliff, river, color, and sky. So today, likewise, the canyon is valued for more than its lessons in geology. Like the embedded river, science might be widening the canyon without deepening it. A new scientific theory could overturn the prevailing geologic interpretation of canyon origins, the earth sciences might find more challenging the Valle Marineris on Mars or the Pacific Ocean's Mariana Trench, but the canyon would still remain a national shrine. It would endure as one of America's sacred places and a site for personal revelation.

Geology was the means by which American society first engaged the canyon as a place of more than passing curiosity, and geology will remain the prism through which it is viewed as a scientific spectacle. *Carving Grand Canyon* reminds us how complex that task actually was. And with his survey Wayne Ranney joins the ranks of those who have made it happen.

—Stephen J. Pyne

Author of *How the Canyon Become Grand: A Brief History* and *Fire on the Rim: A Firefighter's Season at the Grand Canyon*

Preface

I love the Grand Canyon. For me there is no more powerful place on Earth than this huge chasm of stone, water, and light. Since 1975, when I arrived here as a wide-eyed, twenty-one-year-old boy, the canyon and the surrounding landscape have been the recreational, intellectual, and spiritual focus of my life. I am fortunate that my work as a geologist and trail guide has allowed me to repeatedly enter the canyon, where I have developed an intimate relationship with the sweeping ramparts and vast silence of this great, excavated space. I am grateful to know the Grand Canyon so well; its lessons have engendered in me the deepest respect for it and the ideas related to its puzzling evolution. It has served as the perfect classroom to teach geology, but it also has been a teacher and mentor to me. I became a geologist because of the awe-inspiring lessons in earth history it provides. I came of age as a human being in the depths of its salubrious embrace, and in a very real way, I am a child of the Grand Canyon.

I have been privileged to participate in two of the three professional conferences held on how the Colorado River could have carved its Grand Canyon. Convening with my fellow geologists has given me the opportunity to keep up to date with the most current theories about the canyon's formation. In these meetings, the terminology of the geologist prevails, and the discussions overheard are like listening to a foreign language. In spite of my fluency in this language, I am moved to share these scientific findings with a wider audience. Although monographs and websites exist for use by professional geologists, I realize there are very few

Late afternoon in the Grand Canyon of the Colorado River. Photograph by Tom Till places an ordinary lover of the Grand Canyon (if there is such a thing) can go to learn about both the historic and the up-to-date theories related to the canyon's development. Yet most visitors, as they peer into the canyon's colorful and wondrous depths, continue to be fascinated by and wonder about how and when it formed.

After attending my first symposium in 2000 on the origin of the Colorado River, I decided to pursue an idea that preoccupied me since I began taking visitors on backcountry trips into the canyon. Since the 1970s I have taken thousands of people of various backgrounds on multiday backpacks, river trips, or rim tours, providing lectures on the possible origin of the canyon. These lectures, given spontaneously at first, were difficult to organize since I was trying to convey the many nuances that are inherent in such a complex topic. It was unsettling to see the confusion on some faces as I realized halfway through a talk that I had forgotten to mention a basic fact of the river's evolution or the canyon's history.

Compelled by the audience and the Grand Canyon itself, I continued to refine and improve my lectures so that casual visitors could become more familiar with these deeply guarded mysteries. The professional meetings in 2000 and 2010 provided me with a framework of ideas and an incentive to write this book. I have strived to synthesize complex and often competing theories into a more understandable story for readers who may be interested in the story of the canyon but not familiar with the jargon of the geologist. I have received heartfelt encouragement from friends, colleagues, students, and visitors to put my lectures in written form so that a larger audience could share in the fantastic story of the formation of the Colorado River and its Grand Canyon.

Humility, respect, curiosity, unbounded enthusiasm, and ever-increasing circles of earthly beauty and personal exploration are just some of the life lessons that this living gorge has taught me. The Grand Canyon has shaped and guided my life. It is a distinct honor for me to share its story with you, and I hope you will find it as compelling and interesting as I do.

Acknowledgments

A book such as this cannot truly be written by one person—the Grand Canyon is too large a landscape and the topic too complex. I am drawn to tell this story because of the influences of many individuals, and although the words in this book are mine, they have been crafted with the help of professional geologists who know the canyon intimately, and with the help of the many trail companions or rim-bound visitors I am privileged to lead toward the canyon. This seemingly odd conjunction of influences inspires me to tell a complex story in a form that I hope will be approachable for just about anyone. Thank you one and all, whether you shared your research findings with me or asked a "dumb" question. This book benefits from both.

I thank my fellow geologists who perform research in and around the Grand Canyon and share their findings with me. Among these are Sue Beard, George Billingsley, Ron Blakey, Laura Crossey, Ryan Crow, Bill Dickinson, Becky Dorsey, John Douglass, Jim Faulds, Cassie Fenton, Brian Gootee, Carol Hill, Richard Holm, Kyle House, Karl Karlstrom, Jessica Lopez Pearce, Ivo Lucchitta, Kris McDougall, Norman Meek, Andre Potochnik, Jon Spencer, Brian Wernicke, and Dick Young. Bill Breed facilitated my very first river trip in Grand Canyon, and Pete Winn told me my first Grand Canyon geology story. I extend a special thank you to Ivo Lucchitta, friend and mentor, who provided the initial inspiration to me to share the story of the Grand Canyon with non-geologists. I also acknowledge the first geologists to write Grand Canyon origin books for a general audience: N. H. Darton, 1917; Eddie McKee, 1931; and John Maxson, 1962. I hope this work carries on in the spirit and tradition you began.

Palisades of the Desert. Photograph by Mike Buchheit

To the many guests who have taken a river trip, hike, or guided tour with me, I extend heartfelt thanks for your support and belief that I could tell this story. May all those who follow in your footsteps be as interested in this story as you have been. My fellow river and trail guides are too numerous to list individually, but your special knowledge of the canyon helped me to formulate a coherent story for a wider audience. Thank you for your companionship as we have explored the "big ditch" together. I thank the Museum of Northern Arizona Ventures Program and the Grand Canyon Field Institute for continuing to provide me with opportunities to take backpackers and river runners into the canyon.

I thank the Grand Canyon Association (GCA) for the opportunity they have given me to tell this story. My editor, Faith Marcovecchio, was a blessing in the whole process and saved me from a steep writer's "precipice" as I ran off track in composing parts of this second edition. Todd Berger, formerly of GCA, also contributed a welcomed guiding hand to the editorial process. Bronze Black made the colorful maps and diagrams look so good. Bronze, you rock! Stewart Aitchison gave me my first opportunity to become an international guide and encouraged me to put my Grand Canyon origin talks into book form. Thanks to the Frampton, Mistretta, and Hansen-Nelson families whose questions about the Grand Canyon, while hiking the Highline Trail in the Canadian Rockies one fine August day in 2003, inspired the book's summary chapter. Thank you to Carl Bowman, Allyson Mathis, and Johanna Lombard from the National Park Service for their review of the manuscript. Lastly, to my greatest fan and my biggest supporter who always believes in me, love and thanks to my wife, Helen. It was this book that first brought us together, and for that reason alone I will be eternally grateful that this book came to be.

Introduction

The Grand Canyon is a world unto itself. When visitors turn their back to the plateau that ends abruptly at the canyon's edge and descend a Grand Canyon trail, they enter an ethereal world of light and stone. It is a landscape separate and distinct from the one that surrounds it, yet very much coherent and integrated within itself. To some it may appear as if in ruins, perhaps even unfinished, as the jagged cliffs yield ever more boulders to the slopes below. It is probably true that the depth and profile of the canyon have not changed significantly since the end of the last ice age some ten thousand years ago. In many ways, the Grand Canyon is an enigma, but one that ultimately inspires and creates the greatest awe within the souls of viewers.

This canyon is one of our planet's most sublime and spectacular landscapes, yet to this day it defies complete understanding. It is visited by millions of people a year, and not one of them knows precisely how or when it formed. The canyon's birth is shrouded in hazy mystery, cloaked in intrigue, and filled with enigmatic puzzles.

The Grand Canyon is a source of inspiration and enchantment for many who fall under its spell. Photograph by Gary Ladd

That the canyon's origin is a mystery may come as a surprise to those who know that geologists speak with confidence about certain aspects of our planet's ancient history and the life-forms that once lived here. We routinely search the far reaches of our solar system for planetary volcanoes, Martian water, or frozen seas of methane. Yet even today, there remain competing theories about how the Grand Canyon, on our home planet, may have formed. How can one of the most treasured landscapes on Earth defy a unifying theory regarding its formation?

The foremost reason why there isn't one accepted theory is that the Grand Canyon has been shaped largely by erosion. As the Colorado River has continued to deepen its track, other forces of erosion widen the canyon and remove most of the evidence for its early incarnation. As the canyon has become deeper and wider, its history has receded into the shadowy depths, much like the cliffs that enclose it when the sun sinks below the rim. We may never know the intimate details of Grand Canyon's origin simply because so much of the evidence has already been washed downstream and is gone forever.

However, the one aspect of the canyon's character that may be most responsible for the lack of a unifying theory is its immense size with respect to both time and space. The Grand Canyon is simply too big and too complex to be known easily. It has existed through so vast a time and within such a large space that our short, human frame of reference cannot possibly grasp such a monumental puzzle. Its history straddles many distinct geologic events that have created much of our western landscape, adding layer upon layer of complexity and doubt. Just imagine, it is so vast that various parts of the canyon may have formed at completely different times.

To comprehend Grand Canyon's size, we must consider both its length and its depth. Measured along the river, the canyon is 277 miles long. Arguably, there is no other single canyon in the world that approaches its length. A vehicle traveling at the eye-blurring speed of seventy miles an hour along an imaginary highway parallel to the Colorado River would take four hours to drive through the canyon. The greater portion of its length is a mile deep within the earth, with individual sheer escarpments that rise to heights of hundreds of feet or even thousands of feet. It requires phenomenal stamina and endurance to explore this rugged defile, a staying power only a few hardy souls can muster during their lifetimes.

It is Grand Canyon's immense size and rough terrain that has kept most geologists confined to one or both of its two rims, its few trails, or its single river corridor.

The amount of time that the Colorado River has been carving Grand Canyon is the subject of much controversy and debate. Collectively, geologists can only suggest a range of ages that vary between about 70 and 6 million years, although the younger age is historically the most widely accepted. This wide range is testament to how hard it has been for scientists to definitely determine when the canyon formed. And while the rocks enclosing Grand Canyon are hundreds of times older than the earliest events even peripherally related to the river's history, 6 million years is still an incredibly long time to comprehend.

What this means is that we find ourselves at a loss in determining exactly what constitutes the beginning of the Grand Canyon. For example, when the canyon was only five hundred feet deep and only one mile wide, as it must have been at some point early in its history, could we still call

Rivers such as the Colorado and the Little Colorado (pictured here) carry large volumes of eroded sediment, removing evidence of earlier geologic development of the canyon. Photograph by Mike Buchheit

Defining *Grand Canyon* and *Colorado River*

When used in reference to the modern landscape, the labels *Grand Canyon* and *Colorado River* are straightforward, but their use with regard to prior ancestors may present some confusion, since earlier manifestations obviously differed from their modern descendants. For this reason, professional geologists sometimes indicate references to a prior ancestor with modifiers such as *paleo, proto,* or *ancestral,* used interchangeably. As we shall see, there are innumerable iterations of a paleo–Colorado River, and using any single term for an ever-evolving entity implies a single, universally agreed upon ancestor. Therefore, I have resisted using such modifiers in this book to designate older ancestors of either the Grand Canyon or the Colorado River.

However, the non-use of these modifiers need not necessarily confuse the reader. As you take the journey backward and forward through time with me, remember that the modern Colorado River and its Grand Canyon are features that only came into existence between 6 and 5 million years ago. In those parts of the narrative that involve this or a younger age, you will easily know what those labels mean. In instances that discuss older time periods, the terms *Colorado River* or *Grand Canyon* refer to an ancestor of these modern features, even if those features were incrementally different earlier in geologic time. It is my hope that this subtle distinction is easy to note and comprehend.

that the Grand Canyon? What if it was born in this exact place but within higher rock strata that are now completely eroded from this area? What if it was born from a river that went in the opposite direction to that of the Colorado River today? Would that still be the Grand Canyon? As we attempt to understand how and when the Grand Canyon was formed, we should remember that this task also involves *defining* what it is.

The mysteries that surround the origin of the Colorado River in Grand Canyon contribute tremendously to its fascination and charm. Yet they also mean that the story you are about to read has no definitive answer, no "A-ha!" moment at the end. We are frequently left with more questions than answers simply because the river has diligently excavated the traces of its early history, leaving us in amazed bewilderment. But more and more, people ask, "How and when did the Grand Canyon form?" This is one of the few books written recently that focuses specifically on that single question. And although it may not be the last word on the topic, it is my hope that this book will serve as a way for interested people to learn more about Grand Canyon's possible origins.

I present the known story of the canyon's formation in three ways: first, to present ideas on how rivers in general may carve canyons; second, to look chronologically at the many theories that were developed through time by successive generations of geologists; and third, to describe a plausible sequence of geologic events that could create this landscape. The first part tells of the processes that have been at work to create the canyon; the second part takes the reader on a collective journey through the last 150-plus years of scientific inquiry; and the third part suggests a way that the canyon may have formed. The reader will follow these ideas as they were presented, discussed, revised, or discarded by those who spent considerable time and effort studying the canyon's origins. The book is intended to be not only a trip through geologic time but one through human history as well, as humans have attempted to unravel the immense mystery and philosophical depth of this wondrous gorge.

In this book, I do not favor one theory over another. I want the reader to make his or her own informed conclusions. I admit that not all theories are mutually compatible, but I feel every researcher is entitled to be a part of the discussion, and the Grand Canyon is a big place that can easily accommodate a variety of views, if only we accept that all of these views emanate from us, incredibly small but curious beings.

Let us take a journey then—a journey through time along the Colorado River in Grand Canyon. We will witness ancient rivers that were born in the wake of retreating seas, experience crustal uplift as our continent was squeezed and compressed, and discuss ways in which a river may reverse its course or how a canyon is deepened. Let's journey with John Wesley Powell, Clarence Dutton, and a host of modern scientists who remain forever fascinated by the Grand Canyon of the Colorado River.

Detail from *Point Sublime* by William H. Holmes, 1882.

Ideas concerning Grand Canyon's formation are in many ways counterintuitive. It appears to have been carved through a previously uplifted plateau, disregards numerous fault lines, was likely born by a river that once flowed the other way, is possibly quite old or quite young—or both—and is set within a more mature landscape.

The Enigma of Grand Canyon 1

Some people may wonder why there is still so much controversy among geologists regarding the Grand Canyon's origin. Perhaps they suspect that in our modern world, with all of its technologic and scientific advances, questions about the canyon's history have been fully answered. They often seem surprised to learn that geologists still refer to "unresolved problems" associated with Grand Canyon's origin. "Didn't the river carve it?" people invariably ask. The answer is, absolutely yes, and the one fact that every geologist agrees upon is that the Colorado River is responsible for the carving of the Grand Canyon. But deeper questions remain: *How* did the river cut the canyon? *When* did it accomplish its task and by which manner of erosion? Geologists remain perplexed by these more difficult questions and continue to puzzle over the subtle intricacies and lack of meaningful clues about how and when this landscape evolved.

Grand Canyon is somewhat unusual among our national parks because of the lack of an easily defined theory regarding the origin of its landscape. The Grand Tetons are known to be one of the youngest mountain ranges in all of the Rocky Mountain system, having been uplifted above the

The Colorado River slices through the Redwall Limestone deep within the heart of Marble Canyon. Photograph by Wayne Ranney

valley of Jackson Hole within the last 13 to 10 million years. When visitors inquire about the Yellowstone landscape, they are told about the catastrophic volcanic eruptions during the last 2 million years that have left the world's largest concentration of geysers, hot springs, fumaroles, and mud pots in their wake. Yosemite Valley yields evidence of being formed (or at least reshaped) by glaciers during the last ice age. But Grand Canyon's origin remains a mystery with numerous uncertainties regarding its age and mode of formation. Until the turn of this century there were few authoritative sources that people could turn to for a basic understanding of how and when it began.

The geologic evolution of national parks such as Grand Teton, Yellowstone, and Yosemite (shown here) are better understood, and the theories less controversial, than Grand Canyon's. Photograph by Wayne Ranney

Part of the puzzle in understanding the Grand Canyon's origin lies in its location, specifically the placement of the river. Some twenty miles east of Grand Canyon Village, the Colorado River turns sharply ninety degrees from a southern course to a western one and into the heart of the Kaibab upwarp. Under ordinary circumstances, an uplifted plateau acts as a barrier to a river's course, causing it to flow around that barrier through lower ground. But the Colorado River and some of its tributaries do not behave like normal rivers; instead, the Colorado appears to cut through

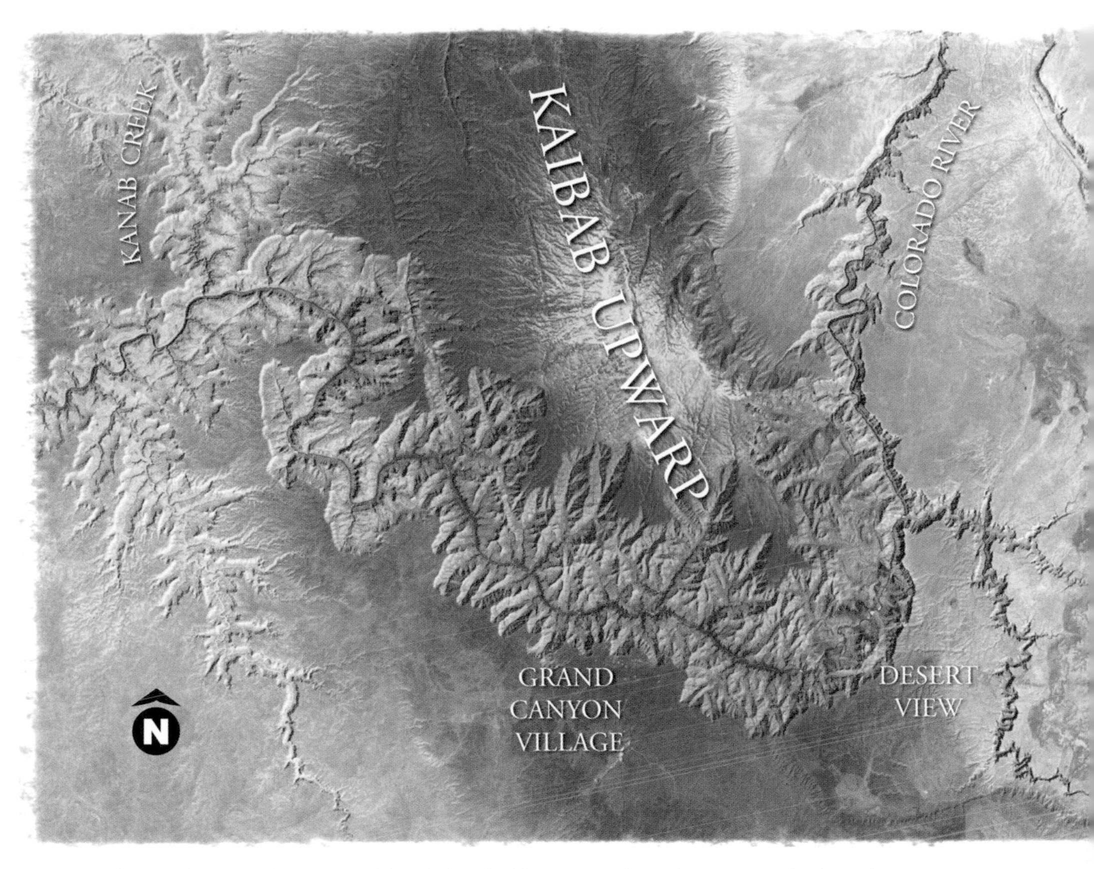

The north-to-south course of the Colorado River turns abruptly west, cutting into the high-elevation, forested Kaibab upwarp (shown in dark green). This odd placement is among the foremost puzzles in understanding the origin of the Colorado River and the Grand Canyon. Landsat image courtesy of U.S. Geological Survey, Southwest Geographic Science Team, Flagstaff Office

an uplifted wall of rock that lies three thousand feet above the adjacent Marble Platform to the east. This odd scenario was the foremost problem recognized by the very first geologists who saw the Grand Canyon. Why does the Colorado River seem to flow into the heart of an uplifted plateau?

Another curiosity with the Colorado River's course is that it often disregards the fault lines that cross its path. Rivers sometimes follow faults where repeated earthquakes have broken and pulverized the ground, creating linear zones of weakened rock. The placement of certain rivers, like the Rio Grande in New Mexico or Oak Creek in Arizona, is more easily accomplished along these lines of shattered rock. Within Grand

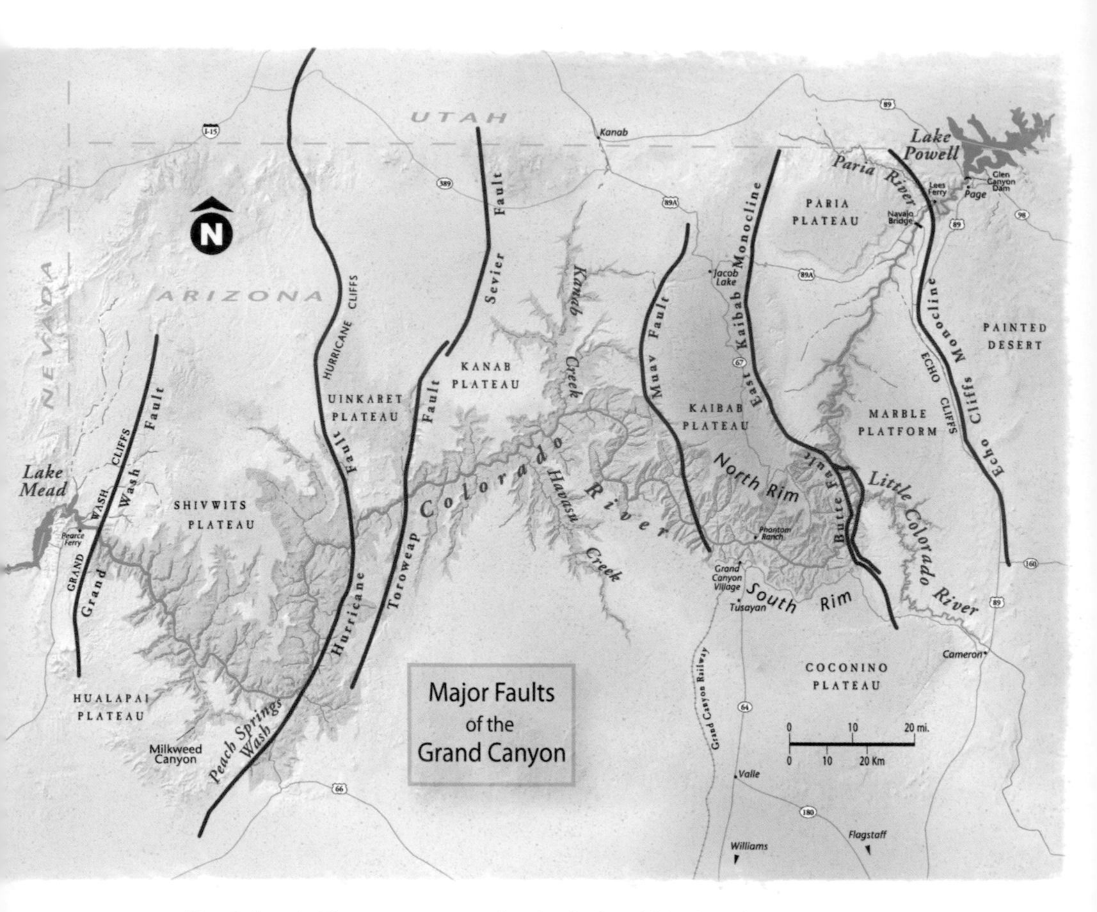

The Colorado River crosses several major faults within Grand Canyon, ignoring these zones of weakness in the landscape and prompting questions about the river's origin.

Canyon, the Colorado River crosses dozens of faults, many of them at right angles to its course, and continues downstream through blocks of strata that are solid and unbroken by faults. Although there is one twenty-five-mile stretch where the river parallels the Hurricane Fault, this is the exception rather than the rule. Why does the Colorado River in Grand Canyon disregard the faults that cross its path, lines that seemingly offer less resistance?

Another fact that begs for explanation is evidence showing that a possible ancestor of the Colorado River flowed in the opposite direction of the modern river today. Certain gravel layers found in western Grand Canyon contain evidence for this, and together with the known geography of the region between about 80 and 30 million years ago, it is very likely that ancient drainage systems here once flowed to the north. This raises additional questions about how the river may have reversed itself and when it might have happened.

One of the most hotly contested matters among scientists is the river's age and thus that of the Grand Canyon. Some geologists see evidence for an old river and canyon at about 70 million years. Others postulate that they are much younger at 6 million years. In the state of Colorado, upstream from the Grand Canyon, the river is certainly older than 10 million years, but in upper Lake Mead, just below the mouth of Grand Canyon, the river cannot be any older than 6 million years. How can a river be of one age in one location but a different age downstream? Novel ideas on how rivers change and evolve through time are the result of this confusing set of circumstances regarding Grand Canyon. We will more closely examine these ideas, which suggest that the Colorado River may have formed from multiple river systems that have been integrated into the single system we see today.

Lastly, the severe depth of the Grand Canyon in relation to the country that encloses it is also a puzzle. Broad, near-featureless plateaus surround the canyon. Early travelers, arriving on horseback or stagecoach, were just as impressed with the remoteness and seemingly endless plateaus that delivered them to the canyon's edge as they were with the color and depth of the great gorge. It may not be readily apparent to the non-geologist that these flat, highly elevated plateaus are worthy of discussion, but they likely formed at an earlier time and by different processes than the deep canyons that dissect them. What sequence of geologic events could have produced such a strikingly different set of landforms so close to one another?

Slickensides are parallel striations that form along faults when two rock bodies move past one another. Photograph by Michael Collier

These perplexing relationships—flow through an elevated plateau, the lack of fault control on the placement of the river, a likely reversal of drainage direction, the canyon's uncertain age, and its setting upon and within a more mature landscape—help scientists to frame the questions needed to gain an understanding of the canyon's origin and evolution. There are numerous possibilities that could answer each of these questions, possibilities that likely amass more queries than answers. Although at first it may seem frustrating to try to grasp such an elusive problem, as geologists dig deeper into these mysteries, their enthusiasm and determination seem to increase as they try to resolve the origins of this world-class landform.

Those curious enough to ask these questions rely on the scientific method to find satisfactory answers. Using this method, a careful observer will frame a question regarding a specific problem; for example, How did the Grand Canyon form? The observer will then propose a hypothesis based on the evidence they find on the landscape. New hypotheses must take into account the findings and constraints from previous theories. Prior theories, however, should not hold back newer ideas simply because they were first. All ideas, old or new, must be tested by scientists to see if they hold up under closer scientific scrutiny. In time, one idea may be shown to hold up more favorably than others. And eventually one may even come to be regarded as an accepted theory. This scientific method has served scientists well in deciphering the complex geologic history of our planet.

Sometimes, however, as is the case with the Grand Canyon, there is not enough evidence to cause everyone to agree on a single solution. Because of the lack of evidence, there may be many possible answers to a single question, and professional disagreements may ensue among geologists. As thoughtful observers, we must satisfy ourselves with the knowledge that we may never be able to fully explain what we see so vividly laid out before our eyes.

Hypothesis or Theory?

In discussing the origin of the Colorado River and Grand Canyon, I make frequent references to certain hypotheses that have been proposed or theories that are accepted. Many times, these terms are misconstrued or conflated with one another, although they are quite different. A hypothesis is a provisional or interim idea that is brought before the scientific community as a proposal in need of further testing. A hypothesis is not casually presented, but has some level of support or data that can back it up. As the larger community of scientists tests any hypothesis, it may stand or fall based on the rigorous testing that ensues. A hypothesis therefore is a proposal that may or may not stand up to closer scrutiny in the long run.

A theory, properly defined, is something quite different. Dr. Lawrence Principe, professor of humanities at Johns Hopkins University, gives a lucid definition of a theory:

> The word theory properly used does not mean a guess or a supposition. Often people contrast theory and fact, as in the dismissive statement, "Oh, that's just a theory." But, that is incorrect. A theory is not something waiting to be proven, to grow up into being a fact. A theory is something far greater than a fact could ever be. A theory is a well-supported, explanatory structure capable of explaining and predicting a range of phenomena. A fact is just an isolated tidbit of knowledge, but a theory organizes facts, concepts, and predictions into a functional, scientific framework. So, for example, the theory of gravity explains and predicts planetary motions, projectile motions, and falling apples. But no one says, "The theory of gravity is just a theory."

Regarding the origin of the Grand Canyon, we might feel challenged to use the word *theory*, since geologic experiments are inherently not reproducible—no one can physically go back in time and test an idea to see how certain landscape features evolved. However, that does not mean that all ideas related to the canyon's origin are necessarily relegated to hypotheses. There are some well-established theories invoked for the processes that likely created the canyon (e.g., headward erosion or basin spillover), and any mystery that is implied involves merely the relative importance that each process might have played. For example, it is possible that multiple processes acted in concert to form the canyon, but that does not render any single process an invalid theory.

Although many of the ideas presented here began as hypotheses, they can rightly be called theories since they have withstood decades of scrutiny—with one caveat. No single theory (as of yet) can answer every question or detail about the canyon's or the river's evolution. Such is the interpretive nature of the science of geology.

The Grand Canyon is carved into a series of elevated plateaus in northern Arizona. Measured along the river, the canyon is 277 miles long and contains numerous subdivisions within it. It is part of the Colorado Plateau geologic province, and five independent factors have acted in concert to produce this stunning landscape.

The Physical Setting 2

If we are to try to make sense of how and when the Grand Canyon formed, we should first familiarize ourselves with the larger landscape in which it is carved. Before people visit the canyon, they might assume that such a remarkable feature is located among other western landmarks such as the spectacular Rocky Mountains or Arizona's rugged Sonoran Desert. However, the Grand Canyon is not found near these other wonders; rather, it is carved into a series of flat, elevated plateaus that, to the visitor, may appear at first to be monotonous or not worthy of notice.

As one approaches the canyon from the south or north, there is no hint as to what lies ahead. Some visitors may even wonder if they are traveling in the right direction. Although an observant traveler may notice a few quick glimpses of the North Rim when approaching from the south on Arizona State Highway 64, most visitors rely on highway mileage signs as evidence they are on the right road. The numerous volcanoes seen just outside of Flagstaff or Williams, Arizona, may provide some visual relief for those who travel with only their destination in mind. But at Valle, a mere thirty miles south of the canyon's edge, the

The Little Colorado River (lower right) enters the Colorado River below Cape Solitude at the south end of Marble Canyon. Photograph by Chuck Lawsen

road drops down into a treeless flatland of grass and sage. Whether approaching the canyon from the north or south, one must travel across broad, seemingly limitless plateaus.

The Grand Canyon has been carved into a series of flat, nearly featureless plateaus, placing two very different landscapes next to one another and perhaps suggesting two different times and styles of erosion. Photograph by Wayne Ranney

On the north side of the river are four plateaus. From east to west they are the Kaibab, Kanab, Uinkaret, and Shivwits Plateaus. On the south side are the Coconino and Hualapai Plateaus. The upper stretch of the Colorado River in Marble Canyon bisects the Marble Platform, which is therefore found on either side of the river in that location. In all there are seven plateaus. What separates the plateaus from one another are the various faults and folds that have cracked and warped the landscape, making it look like an old phonograph record left out in the hot summer sun. As we shall see, some of the boundaries that separate the plateaus are defined by active faults, which have played a major role in determining how the Colorado River has deepened the Grand Canyon.

These plateaus are impressive landscape elements in their own right. As one prominent student of the canyon observed, if the state of Montana had not already co-opted the term, the plateau landscape surrounding the

Grand Canyon could justifiably be called "Big Sky Country." Spectacular cumulus clouds may often be seen trailing away in all directions above the plateaus, the clouds' flat bottoms tracing out the gentle curvature of the earth. The sheer size of the plateaus immediately surrounding the Grand Canyon is remarkable; collectively they are larger than nine different states (the largest of which is Maryland) or the individual countries of Belgium or Armenia. Their elevation varies with geologic structure from a low of about four thousand feet to more than nine thousand feet above sea level.

In the eastern Grand Canyon, which is the part that most visitors see, the Colorado River separates the Kaibab Plateau to the north from the Coconino Plateau to the south. These plateaus may once have been a continuous feature before the carving of the canyon separated them. Both are part of the Kaibab upwarp, an elongate "blister" or dome on the earth's crust that was uplifted slightly higher than the surrounding terrain. When the Colorado River cut the Grand Canyon, it did so not on the crest of the upwarp but to the south of it where the rock strata dip in that direction at about one to two degrees. The human eye does not easily perceive this gentle angle, and some people are surprised to learn that the North Rim (at about eighty-two hundred feet) is twelve hundred feet higher in elevation than the South Rim (at about seven thousand feet).

This southward slope causes more runoff to enter the canyon from the north side, allowing for more erosion in the side streams on the north side of the river. For this reason, the North Rim is eroded away from the river about twice as far as the South Rim. Between its two rims, the Grand Canyon averages ten miles wide and one mile deep. Scientists estimate that the amount of rock excavated from the Grand Canyon is about one thousand cubic miles—a phenomenal amount of material considering that the volume of all river water in the world totals only about five hundred cubic miles. People often ask where all of that dirt has gone; the answer is that farmers are growing carrots and lettuce in it in the Imperial Valley west of Yuma, Arizona. This is where the Colorado River formerly dumped its load of sediment.

Within the Grand Canyon are a number of additional and interesting landforms. Located three-quarters of the way into the canyon is a broad, greenish terrace called the Tonto Platform. It has formed in the eastern half of the canyon where the soft and easily eroded Bright Angel Shale has retreated away from the river at a much faster rate than the underlying

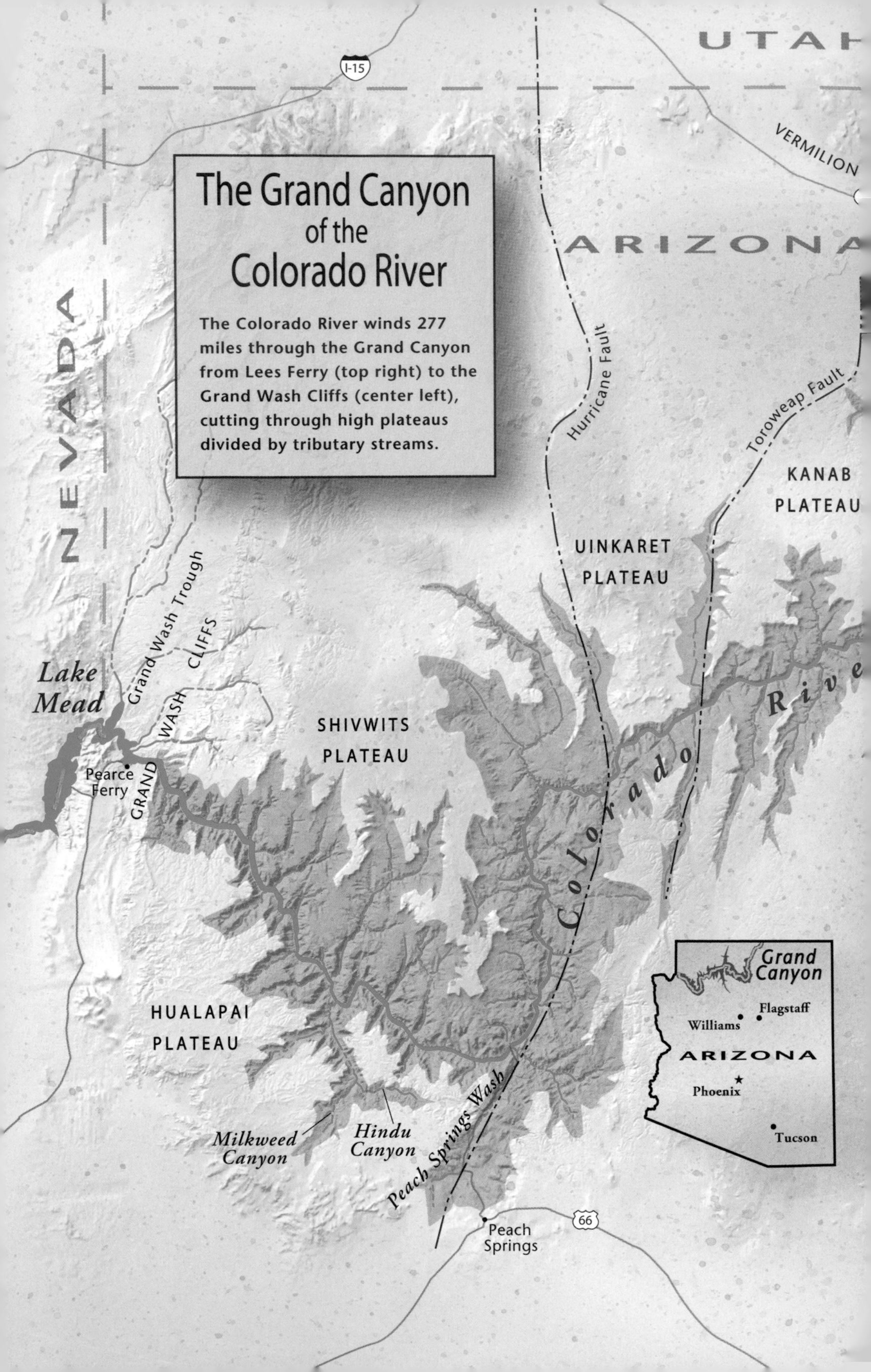

The Grand Canyon of the Colorado River
The Colorado River winds 277 miles through the Grand Canyon from Lees Ferry (top right) to the Grand Wash Cliffs (center left), cutting through high plateaus divided by tributary streams.
UTAH
ARIZONA
NEVADA
I-15
VERMILION
Hurricane Fault
Toroweap Fault
KANAB PLATEAU
UINKARET PLATEAU
Grand Wash Trough
CLIFFS
WASH
GRAND
Lake Mead
Pearce Ferry
SHIVWITS PLATEAU
Colorado River
HUALAPAI PLATEAU
Milkweed Canyon
Hindu Canyon
Peach Springs Wash
Peach Springs
66
Grand Canyon
Flagstaff
Williams
ARIZONA
Phoenix
Tucson

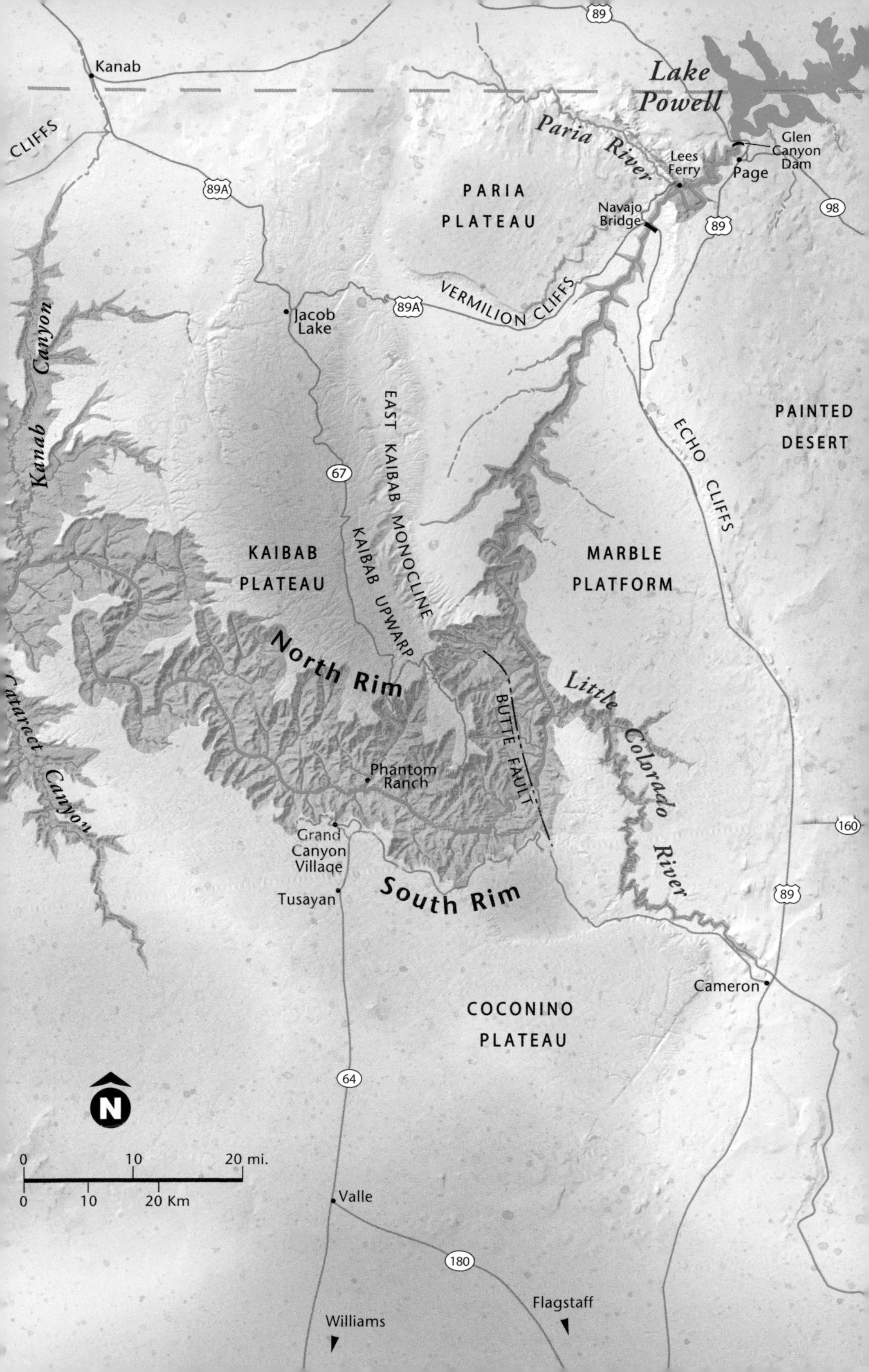

Kanab
Lake Powell
CLIFFS
Paria River
Lees Ferry
Glen Canyon Dam
Page
89
89A
PARIA PLATEAU
Navajo Bridge
98
VERMILION CLIFFS
Jacob Lake
Kanab Canyon
EAST KAIBAB MONOCLINE
67
PAINTED DESERT
ECHO CLIFFS
KAIBAB PLATEAU
KAIBAB UPWARP
MARBLE PLATFORM
North Rim
BUTTE FAULT
Little Colorado River
Cataract Canyon
Phantom Ranch
160
Grand Canyon Village
Tusayan
South Rim
Cameron
COCONINO PLATEAU
64
N
0
10
20 mi.
0
10
20 Km
Valle
180
Flagstaff
Williams

Deep within the canyon, retreat of the soft Bright Angel Shale has formed the Tonto Platform, seen in the bottom third of this photograph. The Colorado River flows from right to left through the Southwest's oldest rocks, forming the Inner Gorge. Photograph by Chuck Lawsen

Tapeats Sandstone and Precambrian crystalline rocks. Beneath the Tonto Platform is the Granite Gorge, cut into relatively hard granite and schist. More than one thousand feet deep in many places, this dark, forbidding slot is the most recent part of the landscape to be revealed by the cutting of the Colorado River. Within the 277-mile length of Grand Canyon are three distinct reaches of the Granite Gorge: Upper, Middle, and Lower. It is the Upper Granite Gorge that is present beneath the hotels on both rims. Lastly, and seen only in the western half of Grand Canyon, is the vermilion-tinged Esplanade Platform. Many aficionados of the canyon consider this landform to be the most graceful example of canyon architecture, since it creates a deeper canyon set within a wider upper canyon. It forms where the soft Hermit Formation retreats away from the underlying harder rocks at a more rapid pace.

Located in the northern part of Arizona (and not in the state of Colorado, as some people mistakenly believe), the Grand Canyon sits isolated from most popular destinations in the Southwest; Las Vegas and

Lake Powell are the closest two and are located at either end of the canyon. When visiting the canyon, travelers must invest a significant amount of time driving or flying over flat terrain. And yet, it is precisely because the canyon is deeply set within these plateaus that it is so intriguing. Plateaus often represent mature landscapes (since their formation involves the lateral removal of the thousands of feet of strata that once covered them), but the extreme depth of the Grand Canyon suggests a more immature landscape that could be described as "under construction," still in the making. When one stands on the Grand Canyon's edge and looks first forward and then backward, two vastly different landscapes are seen, one very much vertical and the other very much horizontal. Combined they tell us how the Grand Canyon regional landscape was formed.

The plateaus, then, are not merely something to put quickly behind us so that we may reach the prize of Grand Canyon. In fact, the plateaus give definition to the canyon itself, and together they combine to form a landscape that is unique on our planet. The plateaus have, in a sense, surrendered any overt beauty they may contain to the obvious glories that lie within the walls of Grand Canyon. Many come to the Grand Canyon

The Esplanade in western Grand Canyon is rarely seen by most visitors. It likely formed by the retreat of the easily eroded Hermit Formation. Photograph by Wayne Ranney

(a few perhaps as landscape "pilgrims") to gaze deeply into this chromatic shrine, but upon arrival they should not forget the stupendous platform that has delivered them to it. In deciphering the story of how the canyon came to be, the plateaus set the stage for how the canyon became so grand.

The seven plateaus surrounding the Grand Canyon are part of a larger physiographic province called the Colorado Plateau. The United States is divided into approximately twenty-six different geographic provinces, defined mainly by their unique landscape characteristics. The Piedmont, Appalachians, Great Plains, and Sierra Nevada are examples of other provinces located in our country. The Colorado Plateau is defined as a region of uplifted but relatively flat-lying sedimentary rocks centered in the Four Corners states of Arizona, New Mexico, Colorado, and Utah. It contains colorful, elevated plateaus that have largely escaped the extreme mountain-building processes that have greatly influenced the development of the adjacent Rocky Mountain and Basin and Range Provinces. The Grand Canyon region is located along the southwestern edge of the Colorado Plateau. In fact, the spectacular and abrupt Grand Wash Cliffs defines the western edges of both Grand Canyon and the Colorado Plateau.

The Rocky Mountain Province is located north and east of the Colorado Plateau. Rocks in this province are varied, being of sedimentary, metamorphic, or igneous origin, but generally elevated much higher than the plateau. To the south and west of the plateau is the Basin and Range Province. This relatively young province formed when North America's crust was stretched and thinned, creating numerous basins that are separated by uplifted ranges. The question of why the Colorado Plateau escaped the intense deformation episodes that greatly altered rocks in the Rocky Mountain and the Basin and Range Provinces is, as we shall see, just now beginning to become clearer. Nevertheless, it is the graceful, flat-lying nature of the colorful sedimentary strata that makes the Colorado Plateau unique.

The Colorado Plateau has an average elevation of between five and six thousand feet and an extreme elevation of more than twelve thousand feet. This means that precipitation comes mostly as snowfall in the winter or as infrequent but intense thunderstorms known as the summer monsoon. The presence of bare rock exposures on the plateau, slickrock in local parlance, allows for catastrophic runoff that coalesces into the

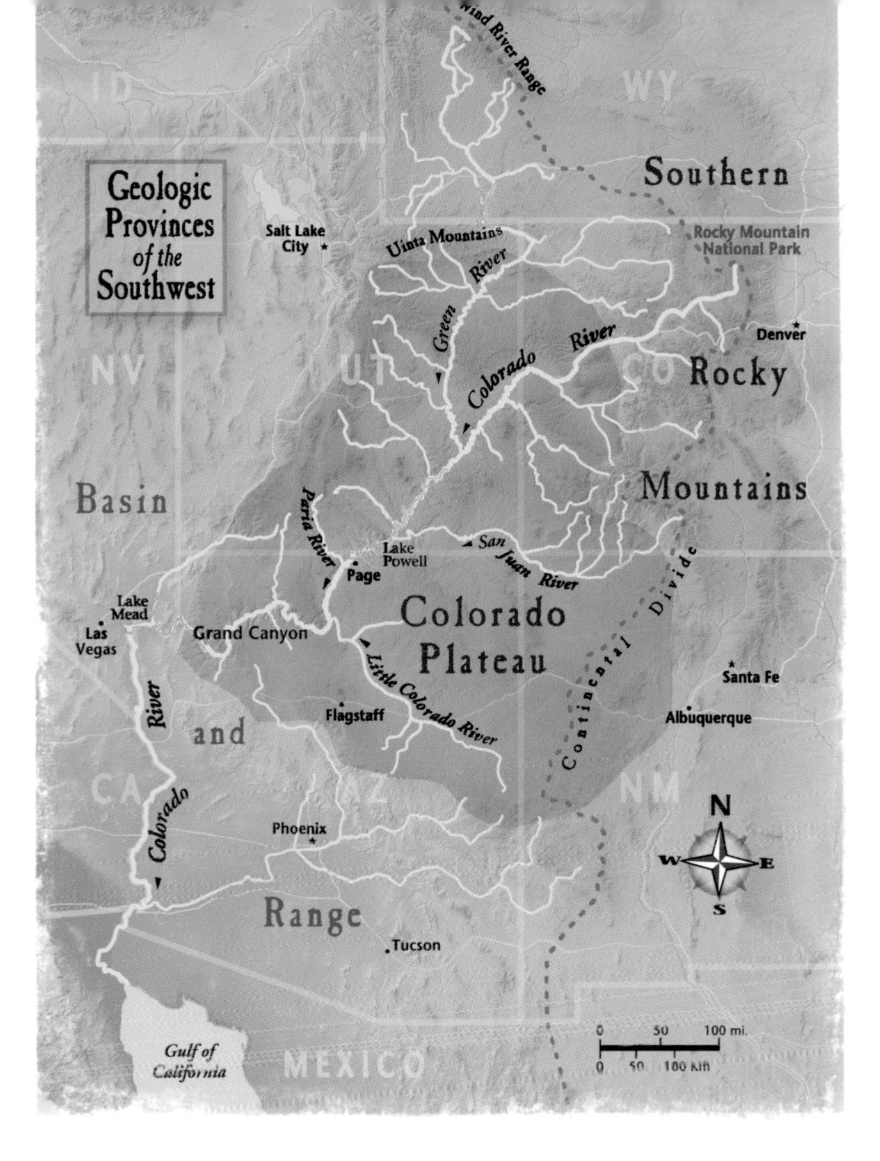

From headwaters to mouth, the Colorado River flows through three distinct geologic provinces.

large rivers that originate in the Rocky Mountains. The Colorado River is the master stream that flows through the center of the plateau and gives the plateau its name. The river begins below the Continental Divide in Rocky Mountain National Park, Colorado, and in the Wind River Range, Wyoming, in the equally contributive Green River. More than 70 percent of the Colorado River's water originates in these headwater mountains. The major tributaries on the plateau are the Gunnison, Yampa, and Dolores Rivers in Colorado; the Escalante and Virgin Rivers in Utah; the

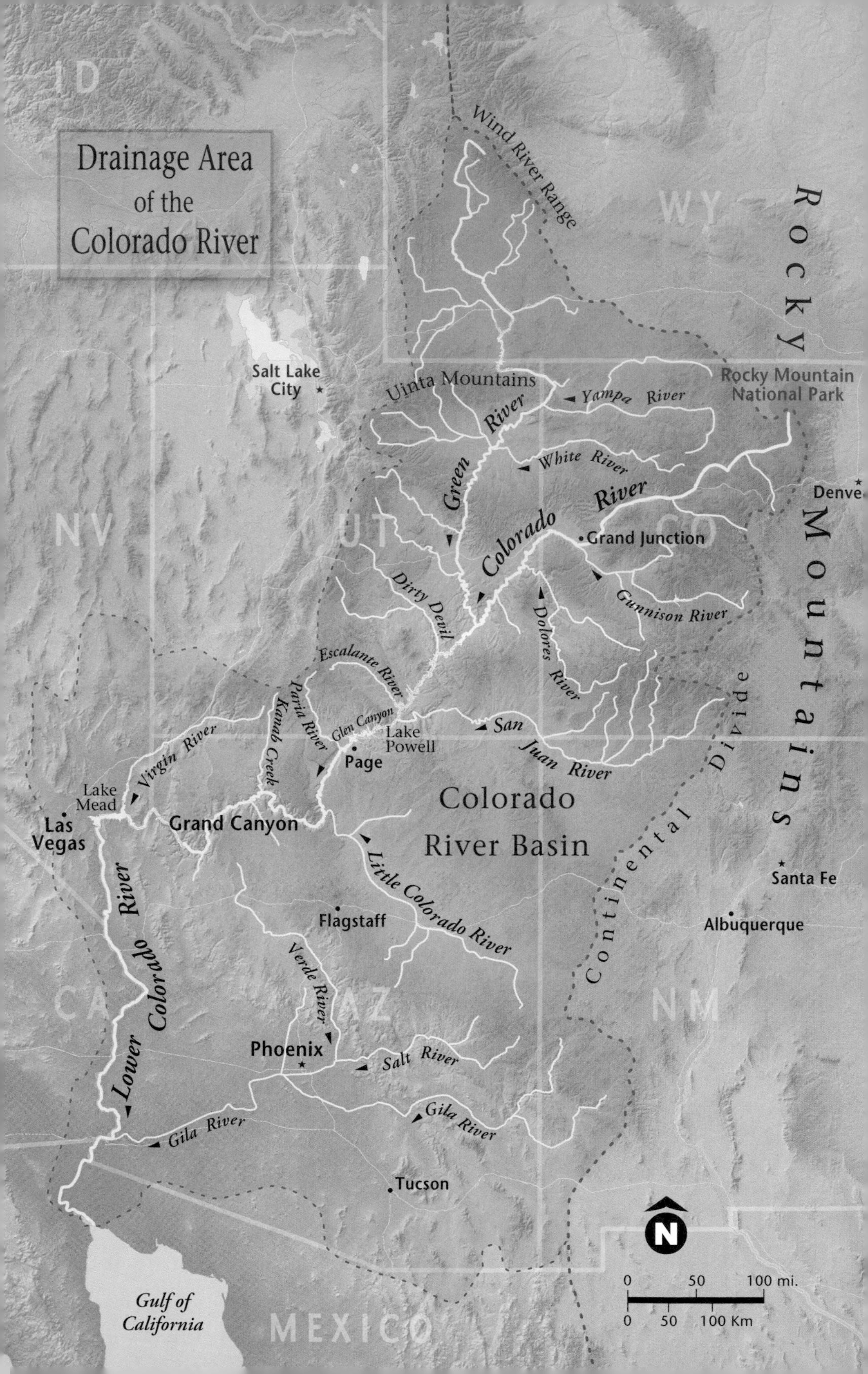

Drainage Area of the Colorado River
ID
WY
NV
UT
CO
CA
AZ
NM
MEXICO
Rocky Mountains
Wind River Range
Uinta Mountains
Rocky Mountain National Park
Salt Lake City
Denver
Grand Junction
Green River
Yampa River
White River
Colorado River
Gunnison River
Dolores River
Dirty Devil
Escalante River
Paria River
Kanab Creek
Glen Canyon
Lake Powell
Page
San Juan River
Virgin River
Lake Mead
Las Vegas
Grand Canyon
Colorado River Basin
Little Colorado River
Continental Divide
Santa Fe
Albuquerque
Flagstaff
Verde River
Phoenix
Salt River
Gila River
Lower Colorado River
Tucson
Gulf of California
N
0 50 100 mi.
0 50 100 Km

San Juan in Colorado, New Mexico, and Utah; and the Little Colorado River in Arizona. The plateau itself may be arid but is coursed by rivers whose branches reach far back into the high backbone of our continent, where precipitation is much greater.

Five independent conditions have acted in concert to produce the unique landscape of the Colorado Plateau. If only one of these were to be removed from the landscape history, the plateau as we know it today would not exist. A few places on our planet approximate the look and feel of the Colorado Plateau, but they are not as extensive in size nor do they contain the same combination of these conditions:

1. A thick stack of stratified rock
2. Vivid and varied color within the strata
3. Widespread, gentle uplift such that the strata remain relatively flat lying
4. The presence of large rivers and their tributaries
5. A modern arid climate

One is immediately struck by the curious combination of large rivers set within an arid climate, and only a handful of other regions on our planet combine these two seemingly mutually exclusive settings. The most obvious are Egypt and its Nile River, along with Iraq and the Tigris and Euphrates Rivers. However, both of these areas lack one of the other conditions necessary to create another Colorado Plateau: high elevation caused by uplift of the earth's crust, upon which those rivers could carve a deep canyon. Runoff from the Rocky Mountains is channeled by the Colorado River across the arid, high-standing plateau that bears its name.

The Colorado Plateau remains unique as a place where these five independent conditions have acted in unison to create this remarkable scenery. When you take an imaginary journey from the lofty Continental Divide in the Rocky Mountains, down into the heart of the Canyonlands and through the Grand Canyon to the Mojave Desert and the Gulf of California, you pass through one of the most interesting and beautiful transects of landscape scenery found anywhere on Earth. This is the stage upon which the story of the Grand Canyon is set.

The Green and Colorado Rivers gather water from numerous tributaries to drain the Colorado River Basin.

Rivers carve canyons when extreme floods roll large boulders along the bed of the river, causing the bedrock channel to abrade and deepen. Coincident with these momentary events, but just as important, are the long-term changes in elevation, river gradient, and climate, as well as the growth and removal of newly formed obstacles.

How Rivers Carve Canyons 3

In our rapidly urbanizing world, most people live their entire lives without ever thinking how rivers like the Colorado carve their canyons. Many visitors to the canyon do not have backgrounds in science or geology, but the sight of it is so spellbinding that their interest is aroused for how the river accomplished its task, and they naturally develop a curiosity for how its profound depth and beauty were attained. They often conjure up ideas that imply a quick or catastrophic origin for the canyon, since it looks to some people like a raw, unfinished work of nature.

It is perhaps natural to think in terms of a catastrophic origin for the Grand Canyon since it does appear like a vast ruin, formed quickly. Yet for many years the simplified scientific explanation was that the canyon probably formed by the slow, inexorable wearing away of the bedrock by the voluminous amounts of sand and silt that once traveled down the Colorado River through the Grand Canyon. Those vast amounts of silty sediment are now trapped behind Glen Canyon Dam, located one hundred miles upstream from the main visitor area.

The Colorado River, seen here looking east from Toroweap Overlook, has carved the spectacular and immense Grand Canyon. Photograph by Tom Brownold

When I was in grade school, I heard that the Colorado River, in all of its brown, silt-laden glory, had slowly and imperceptibly cut through the rock over a long period of geologic time, acting somewhat like sandpaper working very slowly on a piece of wood. The teacher was using the most common explanation for how a gorge so deep could have formed. In my mind, I imagined the tiny particles of silt and sand running over the bedrock channel, polishing it away bit by bit. If this was the manner in which the bedrock wore away, one could live a couple dozen lifetimes and never notice the canyon getting much deeper.

It turns out, however, that very little deepening can be accomplished in this way. For one thing, when geologists conducted studies to determine what the bed of the Colorado River looked like under all that muddy water, they found that the bedrock, defined as the solid, uneroded substrate underlying the river, is rarely exposed to the flow of water. There is up to seventy-five feet or more of sand, gravel, and boulders resting on top of the bedrock beneath the river water. This rocky debris acts as a protective layer that prevents the gritty water (the sandpaper) from wearing away the bedrock (the wood). When this layer of debris was discovered, it became clear that the slow work of muddy water alone could not really have cut the Grand Canyon. Further observations of modern stream processes suggest that rivers only deepen their bedrock channels during relatively rare and intermittent large-scale floods, when huge amounts of large, rocky debris are in motion.

Other factors are at play as well, as I illustrated during an encounter with a visitor at the South Rim. I had just finished a talk to a small group about the Grand Canyon's origin and was looking into the canyon at Yavapai Point. Shortly, a member of the group approached me and asked, "Why don't we have a Grand Canyon in Minnesota?" The question at first seemed odd to me, for as a native westerner I assumed that everyone accepted the fact that deeply dissected landscapes were found primarily in the West and that subdued, mature landscapes were the norm in the Midwest or East. I had never before considered why there wasn't a Grand Canyon in Minnesota. I soon realized, however, that the question relates to one of the most important conditions that must be satisfied for the Colorado River to carve the Grand Canyon: high elevation brought about by uplift.

Left behind as floodwaters receded, these boulders in Marble Canyon represent the type and size of materials that grind into the bedrock channel under flood conditions. Photograph by Gary Ladd

The Mississippi River runs through the Twin Cities of Minneapolis and Saint Paul and is the dominant landscape feature in that large urban area. In my talk I had mentioned that the Mississippi River carried, on average, ten times the volume of water as the Colorado, trying to illustrate the point that as far as North American rivers were concerned the Colorado was not really a big river. This man had astutely surmised that if the Mississippi was ten times the size of the Colorado, then perhaps there should be a great canyon in Minnesota (maybe even ten times bigger). For a moment I was troubled by this line of reasoning, until I realized that I had failed to mention the importance of uplift in the creation of the Grand Canyon. Without the uplift of the landscape, the river could not cut down into it.

How Ancient Floodwater Carved Grand Canyon

Ever since the bed of the Colorado River in Grand Canyon was observed as covered with rocky debris, it has become clear that the river could only deepen its channel when this material was not mantling the bedrock. Geologists make a valid assumption that the material seen on the riverbed today was left there in the waning stages of the last major floods. For the very large boulders (those the size of a large truck or house), these last floods most likely occurred long before humans ever saw the river, perhaps during the height of the last ice age; floods that can move house-sized boulders have never been observed in Grand Canyon. So how do these large-scale floods work to carve canyons?

Flowing water is measured by its discharge, which is the volume of water passing a certain point in a specific amount of time (usually measured in cubic feet per second or cubic meters per second). Doubling the velocity of water in a flood results in a four-fold increase in the size of the particles being transported, and a sixty-four-fold increase in the weight of those particles. When stream velocity and discharge are increased, the stream's competence (the size of material being eroded and transported) and capacity (the total amount of solid load entrained in the floodwater at a given point in time) are also increased dramatically. A river's competence and capacity during floods are many times their levels during low-flow regimes. Essentially, more water at a higher velocity equals a larger carrying capacity and competence to erode bedrock and remove debris.

During huge floods, the capacity and competence of a river may be so large that the entire mantle of debris on the bed of the river is carried away by the floodwater. This exposes the bedrock channel to the rushing water and everything carried with it. This is when huge, car-sized boulders roll forcibly along the channel floor, and as these boulders move and physically pound the bedrock surface, they break off huge chunks of it. In this manner the bedrock channel is ultimately deepened by the intense physical abrasion by the boulders. A single major flood may not accomplish any noticeable evidence of deepening, but five, ten, or twenty of these floods may result in a channel that has been deepened considerably. The waning stages of each flood episode typically leave a new mantle of gravel and boulder debris, which puts the bedrock channel once again into a state of silent preservation beneath muddy water. Short-lived, intense flood episodes are thus responsible for creating the great depth of the Grand Canyon, and these flood episodes may have been more frequent in times past when the climate was wetter and the runoff much greater.

So I turned to this man and responded with another question: "What is the elevation of the Mississippi River in Minneapolis?" We both agreed that it was about seven hundred feet above sea level. I then asked, "How could a one-mile-deep canyon be cut into a landscape that is only seven hundred feet above sea level?" I watched as a light came on for him as he quickly surmised that this would be impossible. Increased elevation of the land, most often the result of uplift, is necessary to carve canyons, and without it there would be no Grand Canyon.

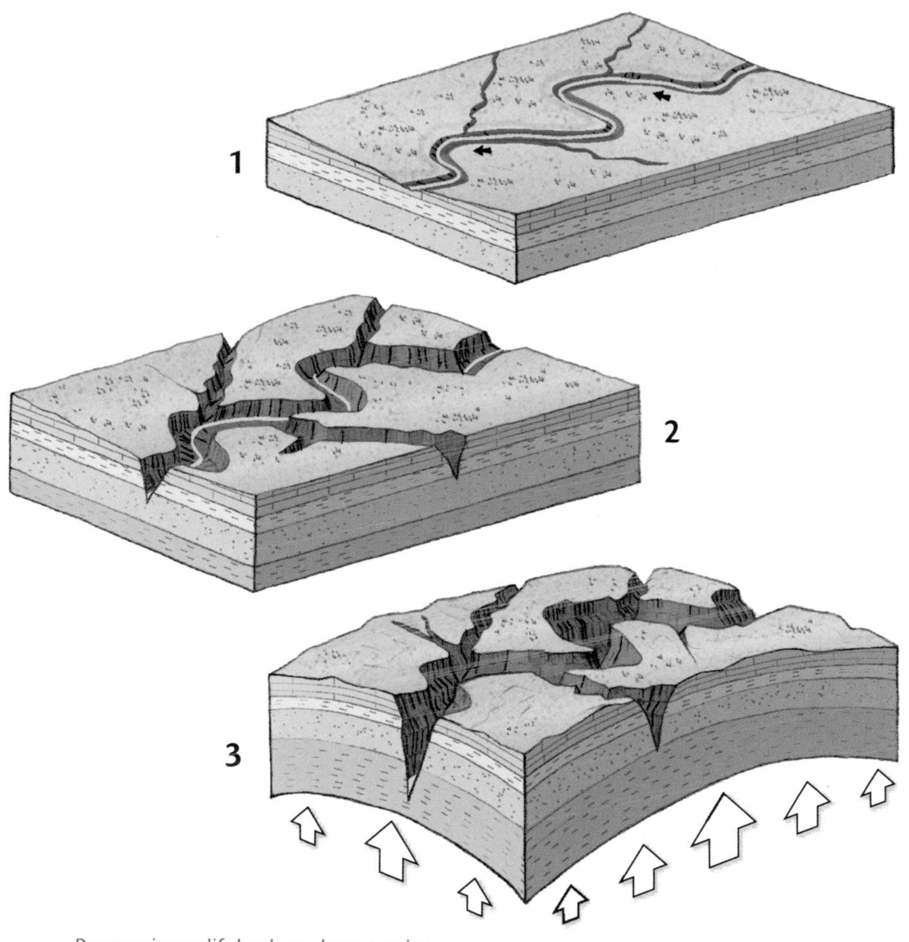

Progressive uplift leads to downcutting.

1 A river flows sluggishly on subdued terrain.

2 As the landscape is uplifted, the river incises accordingly.

3 Differential uplift raises the center part of the landscape at a faster rate than the edges. Canyon cutting keeps pace with the rate and amount of uplift.

Uplift of a landscape can be a difficult concept to grasp because in the course of our short, human life spans, the earth appears to be stable and unchanging. But remember that geologic processes, although often occurring quite slowly to human beings, are always active. During the last 200 million years, the western side of the North American tectonic plate has been compressed as it drifted west relative to the edge of a piece of ocean crust called the Farallon Plate. The result of the collision of these two plates is that the western part of our continent has been crumpled and uplifted, much like a throw rug (North America) is deformed when pushed up against a wall (the Farallon Plate). These lateral pressures are responsible for some of the vertical uplift of the southwestern landscape.

Other forces may be at work as well. Heat is constantly being generated within the earth's interior, and as it tries to escape, it can cause the lithosphere (defined as the earth's crust plus the uppermost rigid part of the mantle) to rise, much like a hot air balloon rises when heated. A variation of this idea is that the lithosphere beneath the Colorado Plateau became delaminated from the crust, rising upward as it became detached from the uppermost mantle. An example of such a delamination might be a piece of Styrofoam that is glued to a sheet of metal beneath it. If the two became unglued while floating in water, the metal sheet would sink and the Styrofoam would rise. Another process that can cause uplift is when the confining weight of overlying rocks is removed by erosion. This type of uplift is called isostatic uplift. The specific cause for uplift of the Colorado Plateau is unresolved, but the plateau has certainly been uplifted since the sea last withdrew from this area. Uplift has raised the strata into a position where erosion can attack it; water is always at work to excavate its way back down to sea level.

The specific timing of plateau uplift is also an important constraint on determining the precise age of the Grand Canyon. While it can be relatively easy to date a volcanic rock (because it contains minerals suitable for use in age dating techniques), the dating of an uplift is much more difficult, since an uplift may not preserve material that can be dated. As we examine the controversies regarding the age of the river and the canyon, we will constantly be referred back to the question of when the uplift of the Colorado Plateau occurred. Geologists know that three distinct periods of uplift occurred throughout the last 70 million years.

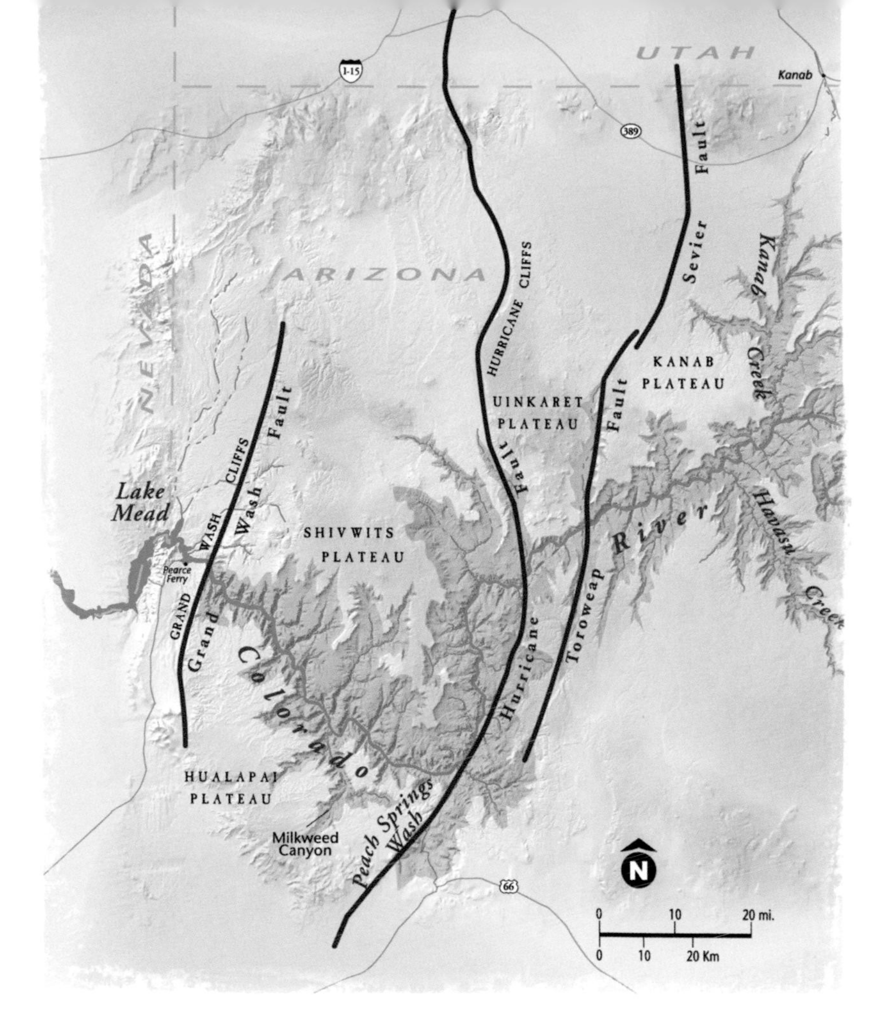

Over the past 3.5 million years, movement along the Hurricane and Toroweap Faults has differentially uplifted the land to the east 1,900 feet while the river's cutting action kept pace.

Some see the uplift early in this time period as being dominant; others argue for later uplift as being most important. Still others argue that what was key was differential uplift, whereby certain portions of the plateau are uplifted or lowered relative to adjacent areas. There is still no consensus on when the majority of the uplift of the Colorado Plateau took place.

However, recognition of recent movements of two faults in the western part of Grand Canyon are providing insights into how differential uplift could have significantly deepened the eastern part of the Grand Canyon in just the last few million years. The Toroweap and Hurricane Faults cross the Colorado River at right angles, and in both instances, these faults caused movement of the earth's crust such that the land on the east, or upstream, side of the fault was raised. This resulted in the uplift of the river channel there, creating a feature on the river called a knickpoint. When knickpoints are formed across a river's channel, they are prone to erosion by the river water, and consequently, the knickpoint gradually migrates upstream. This upstream migration can cause a canyon to become deepened in that direction.

Niagara Falls is the perfect example of knickpoint migration. Before it was altered to produce electricity, the lip of the falls migrated upstream almost four feet per year—an astonishingly fast rate even relative to a human lifetime. The Niagara Gorge downstream from the falls is now seven miles long and has been chiseled down about 160 feet—all in just the last ten thousand years. Although the creation of this knickpoint is not the result of differential uplift along a fault (but rather the glacial erosion that created the Great Lakes), it is a good example of knickpoint migration that deepens a river canyon in the upstream direction.

In the Grand Canyon, knickpoint migration works in much the same way as at Niagara Falls, except that knickpoints in the canyon are formed by the displacement along the Toroweap and Hurricane Faults. The cumulative offset along these faults during the last 3.5 million years is about nineteen hundred feet. This means that the western portion of the canyon has been lowered relative to the eastern side, perhaps causing the western side to escape recent significant deepening. However, upstream from the faults in the eastern part of Grand Canyon, two-fifths of its depth may have been accomplished as this section was uplifted into a position where it could be attacked by river erosion. Relative uplift is a very important concept in understanding how the Grand Canyon evolved.

While uplift is an important component of canyon cutting, climate also plays a crucial role since it determines a river's discharge and thus its velocity, carrying capacity, and competence. Uplift brings rocks up to the elevation where they can be attacked, but it is the elements of climate that do much of the attacking, resulting not only in the deepening

Niagara Falls is perhaps the most recognizable knickpoint on the North American continent. Photograph by Jack Hillers, 1886

of canyons but in their widening as well. Climate and the freeze-and-thaw processes it engenders help to break apart rocks located far from the river channel. When water gets into cracks, it can freeze and expand, prying the cracks open. Then gravity takes over, pulling chunks of rock downslope and eventually widening the canyon. Wind probably does very little to physically wear away rocks in the canyon, but it does transport material that has already been broken into sand-sized grains.

An ancient debris flow deposited nearly seventy-five feet of sediment along Bright Angel Creek. Photograph by Wayne Ranney

During climate episodes with increased rainfall, formations like the Bright Angel Shale can become saturated, resulting in the failure of entire hillsides and creating huge megalandslides within the canyon. These can be volumetrically quite large, with some remarkable examples that have slid down thousands of feet to block the channel of the Colorado River. In a six-mile stretch of canyon between Tapeats Creek and Fishtail Canyon (river mile 134 to 140), huge megalandslides reveal that the canyon is getting significantly wider along this reach. These landslides may sometimes break apart into a watery slush that moves like wet concrete down canyon floors. These are called debris flows, and they, along with megalandslides, have worked to actively widen portions of the Grand Canyon.

In the wetter areas of our globe, especially in tropical settings, increased precipitation and humidity are responsible for the chemical breakdown of rocks, and a flatter, more subdued landscape is generally produced. In these tropical settings, the surrounding hills are lowered at about the same rate that rivers cut into their channels. However, because the Colorado Plateau is located in an arid climate, erosion is quite limited in areas away from the river's edge. Faster erosion occurs in the drainage channels, where infrequent but intense storms coalesce heavy runoff into narrow corridors. In the American Southwest, there is presently much greater vertical dissection, or canyon cutting, along river courses, while the intervening plateaus and mesas are left standing relatively high and dry. About 55 million years ago, the Colorado Plateau was far more humid than it is today, and it was during this time that the broad plateau surfaces may have developed. This may explain how the Grand Canyon was set within a wide, mature plateau landscape.

Understanding climate change through time allows us to better understand how the different parts of the present landscape may have evolved. When the sea last retreated from the Grand Canyon region some 80 million years ago, the American Southwest was located in a humid, subtropical climate belt. The thermal maximum, defined as the hottest and most humid time, occurred about 55 million years ago. Erosion at that time must have been much different from what we see today. Broad planar, or lateral, erosion most likely removed thick sheets of sedimentary strata that used to sit upon the plateau surface above Grand Canyon. Much of these colorful strata are still present to the north in Zion and Bryce Canyon National Parks but may have been removed from the Grand Canyon region during this humid time period. This may be when the Grand Staircase developed on the Colorado Plateau landscape. One of the great canyon geologists, Clarence Dutton, called this cycle of lateral erosion the "Great Denudation."

Slowly through time, as the climate began to cool and become drier, landscape-wide planar erosion probably gave way to more localized vertical erosion along the river courses. Significant aridity in the Southwest

The Grand Staircase and Grand Canyon

The Grand Canyon has been carved one mile deep into the earth's crust, but the rocks within it contain only the bottom one-third of a three-mile-thick section of strata found elsewhere on the Colorado Plateau. These additional strata are thought to have once covered the Grand Canyon (they are still exposed on three sides of it, but were likely removed by lateral erosion before the Grand Canyon was cut). Places to see these strata are in the five national parks in Utah (Zion, Bryce, Capitol Reef, Canyonlands, and Arches), as well as the Navajo and Hopi Indian Reservations in Arizona and the Red Rocks National Conservation Area near Las Vegas, Nevada.

North of Grand Canyon, these strata have been stripped back in such a way as to form a sort of large-scale staircase that steps upward to the north. Grand Canyon's rocks form the base of the staircase, and proceeding north from there, the escarpments climb to increasingly higher (and younger) strata in the stack. Each stair step has a name based on the color of the rock it contains: Chocolate Cliffs, Vermilion Cliffs, White Cliffs, Grey Cliffs, and Pink Cliffs. The rocks at the bottom of the Grand Staircase are very old, up to 1,840 million years, while the rocks at the top are only about 45 million years old. These rocks reveal the many ancient landscapes that used to exist here long before the Grand Canyon came to be. To learn more about these ancient landscapes, see the book *Ancient Landscapes of the Colorado Plateau* by Ron Blakey and Wayne Ranney.

The Grand Staircase north of Grand Canyon at LeFevre Overlook on Highway 89A, with the Vermilion, White, and Pink Cliffs exposed. Photograph by Wayne Ranney

may have begun about 15 million years ago and increased in intensity between 6 and 5 million years ago. This more recent time period may be when the rivers here began cutting the deep canyons. Uplift and climate change most likely worked hand in hand to create the Grand Canyon landscape. Although the specific timing of this uplift remains unknown, the broad aspects of the West's climate regime are becoming more clearly understood, allowing scientists to better interpret how and when the landscape evolved. Many recognizable aspects of the modern, highly incised landscape were most likely beginning to form during the arid conditions of the more recent past.

The related concepts of headward erosion and stream piracy are also important in understanding how some rivers can carve canyons. As we have seen and will discuss further, there are different dates for the age of the Colorado River above and below the Grand Canyon. How can a river be old in one place but young in another? At first this may seem quite inexplicable, but if the river evolved from two separate systems, only to be joined later into an integrated single river, then the puzzle is easily solved. Stream piracy, accomplished by the process of headward erosion, provides us with circumstances that could explain some enigmas concerning the Colorado River. (Headward erosion is more properly termed headward incision, but because the former term is in such common usage with geologists, its use will be continued in this discussion.)

The related concepts of headward erosion and stream capture are important to understanding many theories concerning Grand Canyon's origin.

1 A sluggish river (A) and a steep-gradient river (B) share the landscape.

2 Because of its steeper gradient, the lower river begins to extend its channel upstream toward the lower-gradient river.

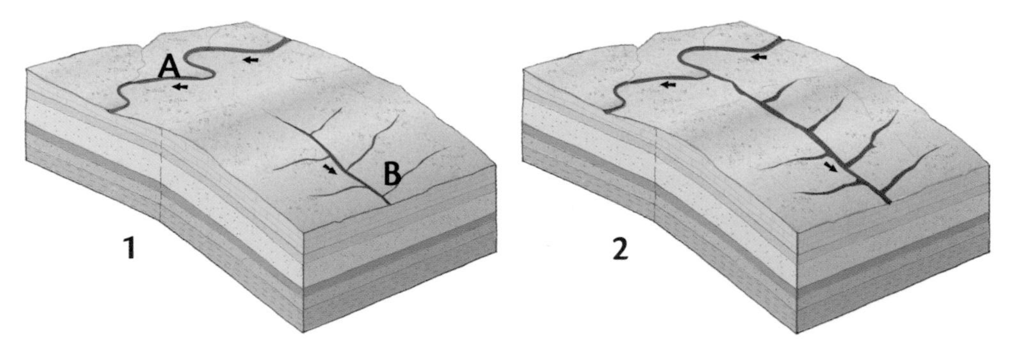

Headward erosion and stream piracy suggest that for two rivers, each located on opposite sides of a drainage divide, the steeper-gradient stream will lengthen its channel in the upstream direction, ultimately capturing runoff that previously belonged to the lower-gradient stream. This occurs because steeper-gradient streams move larger debris in their channels, making them more efficient in attacking bedrock and thus lowering their gradient. Another result of headward erosion is the lengthening of the steep-gradient channel in the upstream direction. When a high-gradient stream lengthens its reach such that it intersects the channel of another stream, it may integrate the upstream portion of that stream into its own. We can envision headward erosion as happening most efficiently on a much more subdued, predissected landscape, long before the deep canyons form. During the integration process, headward erosion may take advantage of the old low-gradient streambeds, which can be thought of as the predetermined perforations in the landscape, dictating where future canyons will be carved. Through time, certain rivers may capture parts of other drainages, dramatically changing the overall configurations of a river system—even causing some sections to experience reversal of flow direction. It is possible that the Colorado River experienced such episodes of headward erosion and stream piracy within the Grand Canyon.

Other processes, such as basin spillover or karst (groundwater cave) collapse, may also act to facilitate the integration of two or more rivers into a single system. Traditionally, geologists tended to minimize or ignore

3 Stream capture occurs when the upper portion of the top river is diverted into the lower river.

4 Ultimately, both parts of the sluggish river become fully integrated with the steep-gradient stream. The courses of both rivers are maintained even as deep dissection follows.

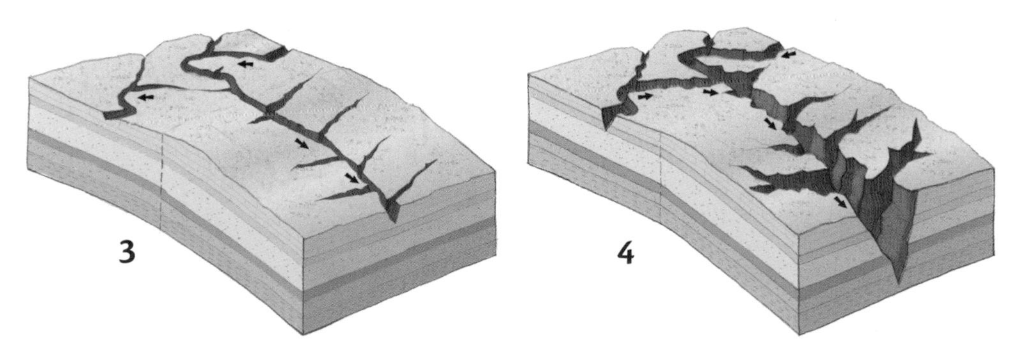

Headward Erosion Today

Headward erosion may be responsible for the integration of ancient separate river systems into the modern Colorado River we see today. Although it's difficult to know the degree to which this process was involved in determining the course of the modern river, a question presents itself: Can we find evidence upon today's landscape where headward erosion may be occurring, which would lead to stream capture and thus add more drainage area to the Colorado River? The answer may be found at the headwaters of the Paria River within Bryce Canyon National Park.

The rim of Bryce Canyon forms a drainage divide between two separate river systems—the Paria River on the east and the East Fork of the Sevier River on the west. A drop of rainwater falling below the rim of Bryce Canyon will flow south and east into the Paria and on to the Colorado River system, but a drop of rain that falls on top of the rim will flow west and north into the Sevier River and the Great Basin (thus never reaching the sea). The East Fork of the Sevier River, which may be a remnant of a former north-directed drainage pattern, has a very low gradient, suggesting it is a more mature stream. The headwaters of the Paria River have very steep gradients indicating that they are young and vigorous. The colorful hoodoos at Bryce Canyon formed as the Paria River's headwall eroded back into the soft and colorful Claron Formation, creating the Paria Amphitheater.

Just two miles from the rim of this amphitheater is the East Fork of the Sevier River. The rim of Bryce Canyon is cutting toward it at the rate of about one foot every sixty-five years—a phenomenally fast rate of change in geologic terms. A quick calculation indicates that continued retreat along this drainage divide would cause the rim of Bryce Canyon to arrive at the Sevier River channel in about 750,000 years. Climate change, relative uplift, or changes in runoff amounts could affect the rate of change. When the rim of Bryce Canyon reaches the Sevier's East Fork, all of the water upstream from that point will be captured and then directed to the southeast and the Colorado River. As headward erosion progresses along the downstream section of the Sevier River, drainage reversal may occur on this stretch of the river and incorporate it as well into the Colorado River system. This present-day landscape relationship lends credence to the idea that the Colorado River may have been cobbled together in times past from separate and distinct river systems.

The Paria Amphitheater below the rim of Bryce Canyon. Photograph by Tom Till

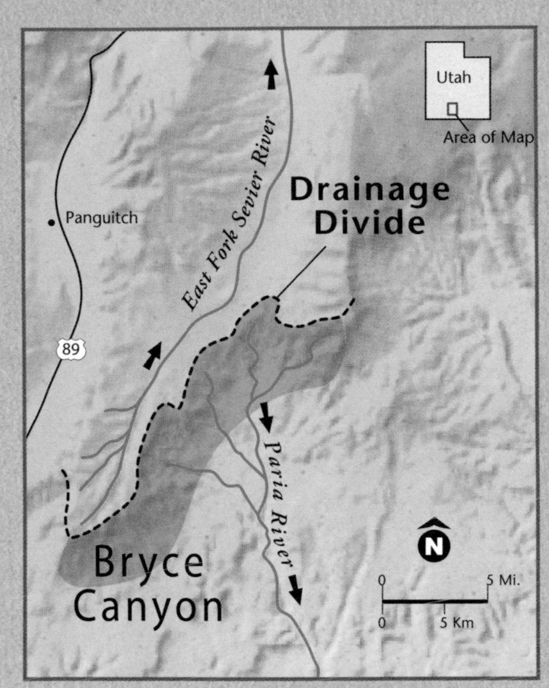

The present-day drainage pattern near Bryce Canyon, Utah, where the rim forms a drainage divide between the steep-gradient Paria River drainage and the much lower-gradient Sevier River.

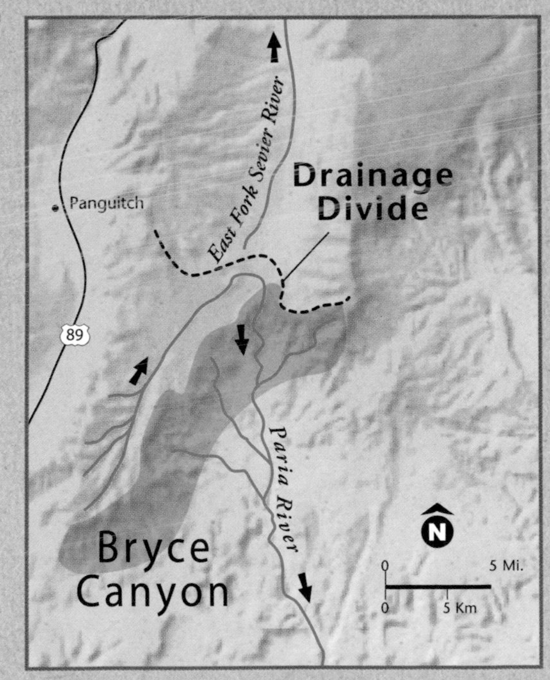

Possible future configuration after headward erosion of the Paria River captures a portion of the Sevier River. Headward erosion will ultimately reverse the flow on the remaining portion of the Sevier River as well, causing the drainage divide to migrate even farther north.

the importance of spillover or karst collapse in an attempt to strengthen the concepts of headward erosion and stream capture. However, a growing number of scientists who specialize in these fields are now presenting evidence that can clarify how these two processes might have facilitated the formation of the Grand Canyon. Basin spillover suggests that basins that were once separated become sequentially filled to their brim with water, causing each to spill across a divide to merge and connect to the next basin below it. The whole process can occur over a few centuries to millennia—not a catastrophic event in a human time frame but still a rather rapid rate to geologists. Karst collapse involves the creation of caves in limestone bedrock that can convey water to the subsurface. This subsurface flow may then influence the direction and configuration of surface runoff if the right conditions are met. Collapse of the underground cave system may lead to the integration of drainages across drainage divides and the development of surface runoff. These two processes have been gaining wider acceptance in recent years for the creation of at least parts of the Colorado drainage, and we will explore them in the next section of the book.

Groundwater is also involved in other ways to help shape the Grand Canyon in a process called sapping, which facilitates canyon widening. When groundwater seeps downward through stratified layers it can readily percolate through aquifer rocks such as sandstone or limestone, but when it encounters a much finer-grained rock, such as shale or mudstone, the water will be horizontally directed and emerges as springs in a cliff wall. Through time, the spring water serves to further weaken the fine-grained strata, such that it begins to undercut the overlying hard layers and they collapse. In this way, the cliff edge gradually retreats as its foundation is weakened. During periods of wetter climate, such as the ice age, groundwater flow in the Grand Canyon may have been much greater than seen today, and the process of sapping was likely a more powerful agent in shaping the canyon than it is today.

One other interesting consideration in understanding the method by which rivers carve canyons is the relationship between the perennial Colorado River and the many dry tributaries that enter it within Grand Canyon. Most of these tributary streams enter the river at grade, meaning at the same level as the main river channel. (There is one notable

Many dry streams, like Pigeon Canyon, enter the Colorado River at grade. Photograph by Scott T. Smith

exception at Deer Creek Falls, but this exception is because of a relatively recent megalandslide that caused the creek's position to change.) How is it that such small tributaries, which have no water in their channels for most of the time, carve canyons just as deep as the Colorado River has carved the Grand Canyon?

The answer to this puzzle is that it is not water that carves canyons but rather the debris that floodwater carries with it. Canyons are carved only during relatively rare flood events, when large amounts of rocky debris physically abrade the normally dry stream channels. Even though these tributaries in Grand Canyon are almost always dry (and thus erosionally inactive), their overall steeper gradients allow them to transport very large material during floods. This means that they carve into the bedrock just

Tapeats Creek (lower left) is a much smaller stream than the Colorado River, but has cut its canyon just as deep as the Grand Canyon. This suggests that it is the gradient of streams and not the amount of water they contain that determines the rate of canyon cutting. Photograph by Larry Lindahl

as readily as the much larger Colorado River. In fact, we could suppose that the tributaries have enough potential power to cut their canyons even deeper than the Colorado cuts its canyon, since their gradients are much steeper. However, they can only cut as deep as the master stream.

Now that we have examined the various ways in which rivers can carve canyons, we can address some misconceptions that still exist regarding the Colorado River. When visitors learn that the canyon is ten miles across, when measured from El Tovar Hotel on the South Rim to the Grand Canyon Lodge on the North Rim, many may assume that the Colorado River must have been much wider at one time than it is today. In fact, the river has probably always been about as wide as it is now, about three hundred feet on average, throughout its time within the walls of Grand Canyon. During the last ice age, when glaciers were repeatedly melting in the Rocky Mountains, the river's volume may have varied quite significantly, but never was the river even marginally close to being ten miles wide. We can surmise this from the fact that nowhere on Earth today is there a river even remotely that wide.

The canyon's great width is the result of tributaries to the Colorado River actively incising their channels, forcing the rim on either side to become undercut and retreat away from the main river channel. Once the Colorado cuts into the rock layers, the tributaries go to work to keep pace with the deepening main gorge. All of these open cuts become susceptible to the other forces of erosion such as freezing and thawing, undercutting, weathering, and gravity, all of which tend to eat away at the canyon walls. While it is true that the layers at the top of the canyon have been exposed to erosion the longest, probably more important in determining the canyon's width is the faster rate at which underlying soft rocks such as shale and mudstone erode, thus undercutting and collapsing the harder layers, for example limestone and sandstone, that overlay them. This is most likely why the canyon is wider at the top. Differential erosion rates of the various strata are the key to understanding how the canyon is widened.

Within the Grand Canyon, the Colorado River has cut through about twenty-three different sedimentary rock formations, as well as through different kinds of metamorphic and igneous rocks near the bottom in the Inner Gorge. Each formation has a certain resistance to erosion based on its rock type, hardness, or density; for instance, a granite or schist layer is much more resistant to erosion than a shale layer. Many scientists believe that when the river is cutting through granite, its rate of downcutting will be slower than when it cuts through shale.

However, there are overriding phenomena that diminish the expected result. Canyon cutting may occur within discrete periods of time that involve the larger and more important parameters of active uplift, climate change, or increased runoff. When these conditions are active, a river such as the Colorado may not cut any faster through soft rock than hard rock, at least with respect to deepening. In other words, it may not matter to any great degree whether the river is cutting through shale or granite; if uplift, climate change, or increased runoff are in play, the channel will become deepened. Rock type does play a significant role in determining the overall width and profile of the canyon walls (such as with sapping), but other factors may have more of an influence in determining the depth of a canyon than rock hardness.

These observations were made clearer when geologists looked at the various lava flows that once blocked the Colorado's path. At least thirteen

How Rivers Are Placed upon a Landscape

Streams find their way across a landscape because water responds to the force of gravity by seeking the lowest path through varied topography. However, erosion has been working for at least 70 million years on the Colorado Plateau, and its effects have sometimes left a confusing set of relationships on the landscape, with some rivers seeming to flow toward or through areas with high topography. For this reason, it is instructive to know some of the ways that rivers can be initially positioned on a landscape.

A stream that originates by flowing down a tilted surface is called a consequent stream. If this stream maintains its course but later flows on an eroded lower surface, it becomes a resequent stream. In opposition to this, a river that flows against the dip of stratified beds is called obsequent. Geologists emphasize that obsequent streams likely started out as consequent (or resequent) streams that later reversed their flow direction. A stream that is positioned in a strike valley (formed along a belt of easily eroded soft rock) is called a subsequent stream.

If a river's course is laid down upon a subdued surface, for example the bed of a recently desiccated lake, and later cuts down and encounters a buried, preexisting topography, it is termed a superposed river. A superposed river can eventually cut down into the preexisting topography no matter how resistant those rocks are. If a river begins on a subdued landscape that is differentially uplifted slowly around it, such that the river's course is not deflected away from the rising structures, it is termed antecedent. The Colorado River system may display evidence for each of these variations in different segments of its length.

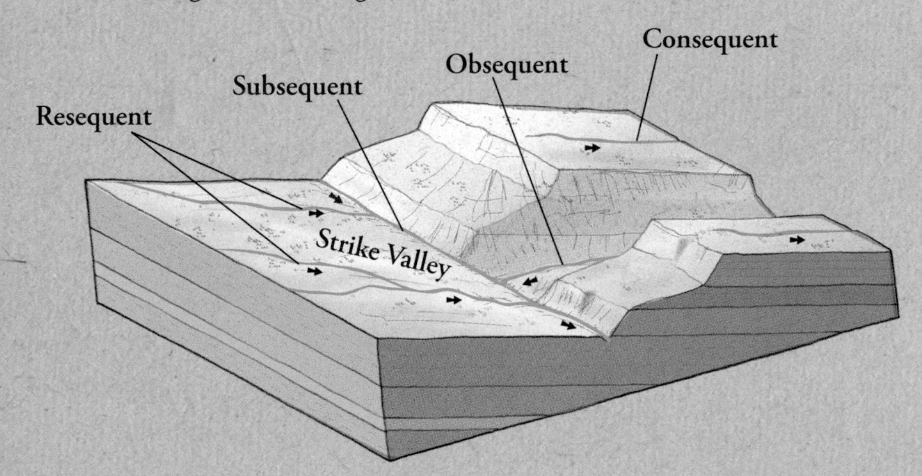

Glen Canyon Dam, located thirteen miles upstream from the head of the Grand Canyon, temporarily obstructs the Colorado River. Photograph by Stewart Aitchison

lava dams have existed within the canyon during the past seven hundred thousand years, and although some of them were quite tall or long, they were destroyed and built again by subsequent lava flows within a relatively short period of time. This shows that no matter what type of obstacle was put in the river's path (and basalt lava can be quite hard and dense), it was removed relatively quickly, regardless of its resistance to erosion. True, a single lava dam may have been inherently unstable from the start since lava was either positioned on top of loose river gravels or quenched quickly in cold river water (or both), therefore causing a dam to collapse catastrophically shortly after it formed. Although one of these dams is documented to have been more than two thousand feet high, and another more than eighty-four miles long, each of these obstacles was removed relatively quickly before the next dam was emplaced.

As a geologist, I am often asked about the effect that Glen Canyon Dam may have on the rate of deepening of the Grand Canyon. The line of reasoning is that the dam has tamed the wild river and its springtime floods, thus stopping the deepening that must have occurred prior to the dam's construction. Certainly, the dam does hold back any large floods and the sedimentary debris that would have previously roared through Grand Canyon. In the short term, Glen Canyon Dam has greatly affected the sediment load and the ecology along beaches within Grand Canyon.

With respect to geologic time, however, Glen Canyon Dam is a temporary feature and will be obsolete within only a few hundred years. Even if we use the youngest age for Grand Canyon ascribed by some geologists, 6 million years, the life span of Glen Canyon Dam pales in comparison. We must remember that the Grand Canyon's evolution has most likely proceeded in fits and starts, and that, even prior to humans' meddling, there were significant periods of time when the Grand Canyon was not being actively incised. The length of these natural breaks from canyon cutting far surpass the time that Glen Canyon Dam will be in existence. Perhaps the river will not even note that a barrier of such short duration as Glen Canyon Dam even existed. Glen Canyon Dam is such a temporary feature in geologic time that it has no effect on retarding the overall rate of cutting of the Grand Canyon.

And so, the picture that is beginning to emerge is that the Colorado River may have cut the Grand Canyon in episodic pulses of active erosion and deepening, interspersed with long periods of quiescence and stability when the canyon remained static. This concept is similar to the one of punctuated equilibrium developed by evolutionary biologists for the descent of species to explain how speciation in animals and plants occurs slowly most of time but then rapidly progresses when some equilibrium is altered by a natural force such as climate change. In other words, certain portions of the canyon's history may include periods of stability or equilibrium during which little to no deepening or widening of the canyon occurs. Those periods are disrupted episodically by periods of great change resulting from increased uplift, climate change, drainage integration, or runoff amounts.

The Inner Gorge of the Grand Canyon is a spectacular example of canyon cutting on a grand scale. Photograph by Gary Ladd

Theories on the origin of the Colorado River and Grand Canyon have been proposed for more than 150 years. As expected, these have evolved through time, yet a single coherent answer is still elusive even though advances in our understanding have been made. Some of the questions that originated in the nineteenth century remain unanswered in the twenty-first century.

History of Geologic Ideas 4

We now turn our attention to the evolution of ideas regarding the origin of Grand Canyon and the Colorado River. Since the first scientists set eyes on the canyon in the middle of the nineteenth century, myriad theories have been put forth as men and women have grappled to understand the canyon's formation. This chapter organizes those ideas into four parts: early ideas from the nineteenth century, alternate theories of the early twentieth century, more modern concepts of the late twentieth century, and theories from two major symposia held in the twenty-first century. Nearly all of the historic theories that were proposed are included here, as well as the major themes that provoke debate.

Note: The dates appearing after the geologists' names refer to the publication dates of their theories and not the dates of their expeditions or research years. Specific references for these are listed in the bibliography. Note that the quotes provided on various theories are key statements taken from much longer treatments of the subject. While they serve to summarize a key idea presented, longer lines of reasoning and more complex and involved data support the ideas contained in the quotes.

The Grand Canyon at sunset. Photograph by Chuck Lawsen

LATE NINETEENTH CENTURY

Fluvialism

Courtesy of USGS

John Strong Newberry, 1861

A common misconception among students of the Grand Canyon is that John Wesley Powell was the first geologist to view it and attempt to explain how it came to be. In fact, that honor goes to John Strong Newberry, who accompanied Lt. Joseph Christmas Ives on his pioneering exploration of the lower Colorado River in 1857–58. Lieutenant Ives has been immortalized in the Grand Canyon region as the man who least understood the significance of the landscape he was instructed to explore. A quotation from his report to the Congress of the United States, often recalled today by Grand Canyon enthusiasts, bears repeating.

> *The region . . . is, of course, altogether valueless. It can be approached only from the south, and after entering it there is nothing to do but leave. Ours has been the first, and will doubtless be the last party of whites to visit this profitless locality. It seems intended by nature, that the Colorado River, along the greater portion of its lonely and majestic way, shall be forever unvisited and undisturbed.*

From a modern perspective, he could not have been more wrong about the canyon being valueless. Today 4.5 million people visit this "profitless locality" each year, with 40 percent of them from beyond the borders of the United States. Rather than scoff at Lieutenant Ives's lack of vision (for in fairness to him he did pen a few complimentary passages about the canyon landscape in the same report), it is helpful to understand the point of reference from which he was speaking. He hailed from New Hampshire and surely held the common belief in those days that a place as dry and highly dissected as the Grand Canyon could never amount to anything useful for our young nation. Although his primary objective was to determine whether the Colorado River could be used to supply military troops engaged in the Mormon campaign, he was likely predisposed to report on the presence or absence of arable land that could be settled and cultivated. Knowing his nineteenth-century bias makes it easier for us to understand why he wasn't impressed. It would take a different, less agrarian orientation to understand that perhaps the Big Cañon, as he called it, would have some redeeming value in the future.

Ironically, that point of reference would come from within his very own exploration party. It was John Strong Newberry who looked upon the same rugged terrain as Ives and saw or sensed a much different landscape. He was trained both as a physician and as a geologist, and as such was prepared to accept with greater appreciation the spectacular natural feature that their party chanced upon. He wrote these words about the Colorado Plateau region shortly after one of his later expeditions north of Grand Canyon:

> *Though valueless to the agriculturalist, dreaded and shunned by the emigrant, the miner, and even the adventurous trapper, the Colorado Plateau is to the geologist a paradise. [It has the] most splendid exposure of stratified rocks that there is in the world.*

Although Professor Newberry struggled alongside Ives in futile attempts to find reliable sources of water for themselves and their horses, he was capable of seeing that the landscape before them was truly unique on Earth. He made the first elementary but ultimately necessary observation that it was erosion by running water that had cut this great chasm:

> *Having this question constantly in mind, and examining, with all possible care, the structure of the great cañon which we enter, I everywhere found evidence of the exclusive action of water in their formation. The opposite sides of the deepest chasm showed perfect correspondence of stratification, conforming to the general dip, and nowhere displacement; and the bottom rock, so often dry and bare, was perhaps deeply eroded, but continuous from side to side.*

In this powerful passage, we see how a trained eye can observe the evidence, without preconception or bias, and make the most fundamental interpretation for what shaped the landscape. Newberry noticed that the sequence of strata on either side of the river was not broken or offset. Thus, the canyon could not have formed from some profound fissure or fault that ripped open the earth, only later to be occupied by the Colorado River. Everything in his view was the result of water cutting down into

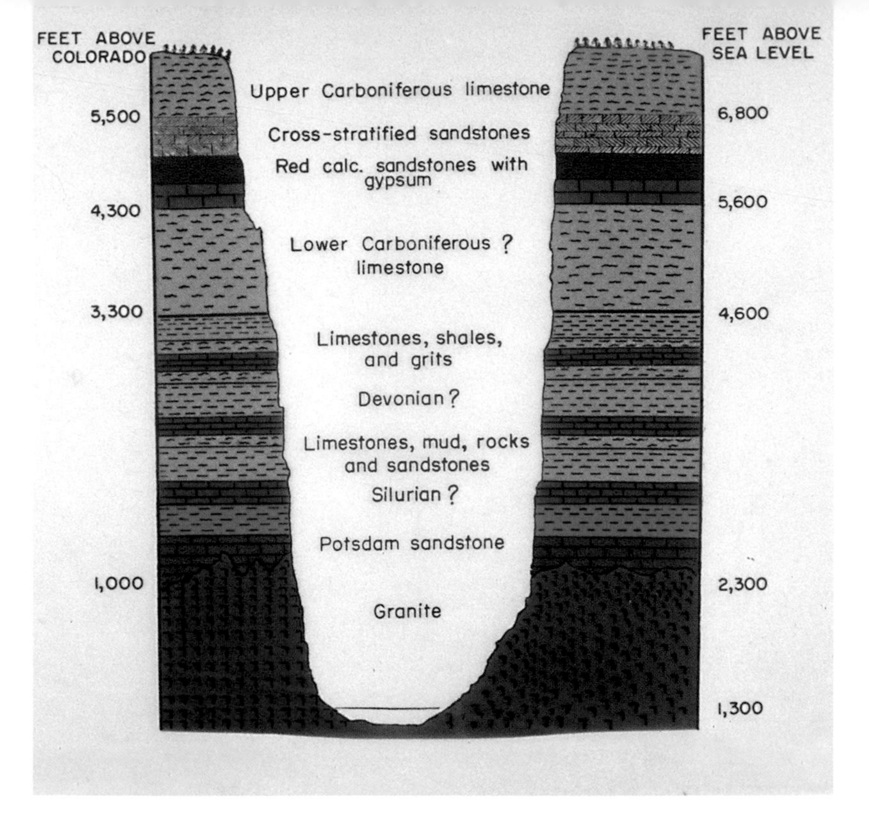

John S. Newberry's original stratigraphic section of the Grand Canyon. Courtesy of USGS

previously unopened earth, even though the side canyons were "dry and bare." This basic observation is the first and foremost lesson that the Grand Canyon reveals to us and remains such to this day.

Newberry also realized how unique the Grand Canyon landscape was with respect to the earth's increasingly known geography, and his observations here spawned a new subdiscipline in geology called fluvialism, a branch that deals with how rivers help to shape landscapes.

> *[These canyons] belong to a vast system of erosion, and are wholly due to the action of water. Probably nowhere in the world has the action of this agent produced results so surprising as regards their magnitude and their peculiar character.*

A positive outcome of Newberry's participation in the Ives expedition is that he made the first stratigraphic column of the Grand Canyon (the first map and hand-drawn sketches of it were also included in the report to Congress). For all of their hardship in the winter of 1858, the expedition was, geologically speaking, a huge success. How strange it seems that

Newberry and Ives would react so differently to the landscape they were charged to explore. That they traveled together, slept on the same cold, hard ground, and suffered from the ever-present lack of reliable water sources, proves that one's perspective makes a great deal of difference regarding adversity.

Antecedence

Courtesy of USGS

John Wesley Powell, 1875

Eleven years would pass before the next geologist laid eyes on the Grand Canyon, and whereas Newberry had the advantage of seeing the Big Cañon from the rim, John Wesley Powell would benefit by observing it from the river. Having spent a few summers conducting scientific research in the Rocky Mountains of the Colorado Territory, Powell's gaze was ever focused from those lofty heights toward the vast and unexplored heartland of the Colorado Plateau. He was determined to make a thorough investigation of what seemed to be an interesting geography.

In May 1869, he set out from Green River, Wyoming, with nine men in four wooden boats and began a 101-day descent of the Green and Colorado Rivers. At the beginning of the trip, Powell had a freshness of spirit that fostered creative scientific thinking about how the course of the Green River had positioned itself with respect to the Uinta Mountains, located in northeast Utah. By the time he reached Grand Canyon, he was weary, nearly out of food, at odds with some of his men, and concerned about the treacherous rapids that choked the river. Because of these problems, his thoughts on the Grand Canyon's origin were made only by inference from his more detailed observations on the Green River, as it cut through the eastern flank of the Uinta Mountains.

In his *Exploration of the Colorado River of the West and Its Tributaries*, published in 1875, Powell agreed with Newberry that erosion by water and not preexisting fissures accounted for the dissection of the plateau country. However, he took Newberry's idea and added to it by suggesting a way in which this erosion had commenced. Powell was struck by the curious fact that the Green River ignored and bypassed the open valleys to the east, only to turn headlong south into a solid bedrock canyon,

An observer looking upstream along the Green River sees a stream that meanders through open valleys, but upon turning 180 degrees from that same vantage point, sees the river plunge headlong through the Gates of Lodore. Photographs by Wayne Ranney

the Canyon of Lodore, on the east flank of the uplifted Uinta Mountains. Common sense dictates that rivers should flow in low valleys and around high mountains, which normally act as barriers to rivers. Yet the Green River did not follow this simple pattern.

To a person studying the physical geography of this country, without a knowledge of its geology, it would seem very strange that the river should cut through the mountains, when, apparently, it might have passed around them to the east, through valleys, for there are such along the north side of the Uintas, extending to the east, where the mountains are degraded to hills, and, passing around these, there are other valleys, extending to the Green, on the south side of the range. Then, why did the river run through the mountains?

The river's relation to the mountains didn't make sense, yet with his geologic training Powell postulated an idea that could explain this rather odd arrangement:

Again, the question returns to us, why did not the stream turn around this great obstruction, rather than pass through it? The answer is that the river had the right of way; in other words, it was running [before] the mountains were formed; not before the rocks of which the mountains are composed were deposited, but before the formations were folded, so as to make a mountain range.

This was a phenomenal leap of insight, not only regarding the immediate question of the evolution of the Colorado River system but also for the science of geology in general. Geologists are forever attracted to confusing sets of relationships (such as that between a river and the landscape upon which it is set) and disposed to offer suggestions for how they came to be. Powell's idea was so new that he needed to choose a new term to describe it: *antecedence*, meaning the river's path predates the uplifted features through which it flows.

I have endeavored above to explain the relation of the valleys of the Uinta Mountains to the stratigraphy, or structural geology, of the region, and, further, to state the conclusion reached, that the drainage was

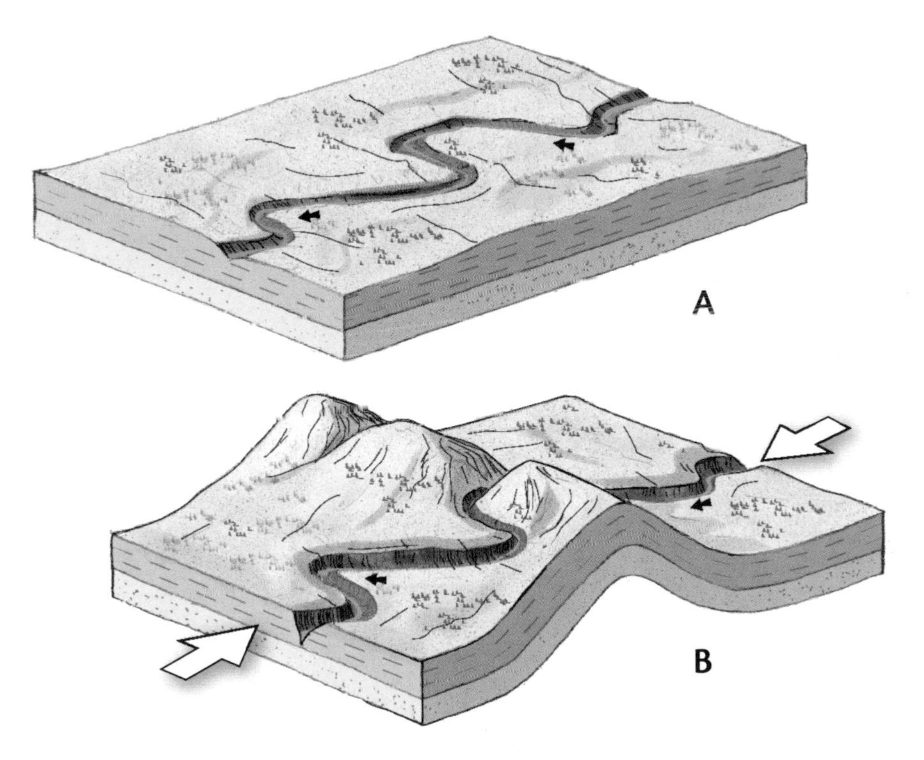

A An antecedent river flows across a very subdued terrain.

B As the terrain is slowly uplifted, the river cuts a channel through the fold to maintain its course.

> *established antecedent to the corrugation or displacement of the beds by faulting and folding. I propose to call such valleys . . . antecedent valleys.*

To understand antecedence, we can read Powell's own explanation of it.

> *We may say, then, that the river did not cut its way down through the mountains, from a height many thousands of feet above its present site, but . . . it cleared away the obstruction by cutting a cañon, [as] the walls were thus elevated on either side. The river preserved its level, but the mountains were lifted up; as the saw revolves on a fixed pivot, while the log through which it cuts is moved along. The river was the saw which cut the mountains in two.*

Viewed this way, the river retained its course even as the land rose up around it, albeit slowly enough so that the river's course was not deflected by the rising mass of rock. Later refinements would challenge the theory of antecedence, but Powell provided a clever and thoughtful explanation for how the odd relationship between the Green River and the landscape that enclosed it came to be.

Even so, there was another possibility that could explain what was seen on the Green River, and Powell was aware of it. After Powell's two river trips but before the publication of his report in 1875, a geologist named Archibald Marvine mapped the Rocky Mountain region in Colorado. In Marvine's report of June 1874, he interpreted a different but entirely plausible sequence of events for the development of the landscape. Marvine suggested that the streams there had originated on top of young sedimentary material, which covered and buried an older, mountainous topography. These streams eventually cut through the flat sedimentary cover and chiseled their way down into this buried topography. This is a subtle but significant difference from antecedence. In this theory, the river remains the saw but moves down to a buried and stationary mountain mass; in other words, uplift is not a part of the process.

Marvine's interpretations seem overshadowed by Powell's rising fame as a successful explorer-*cum*–Washington bureaucrat (he was the second director of the U.S. Geological Survey and the first director of the Bureau of Ethnology), but Powell himself thought enough of Marvine's hypothesis

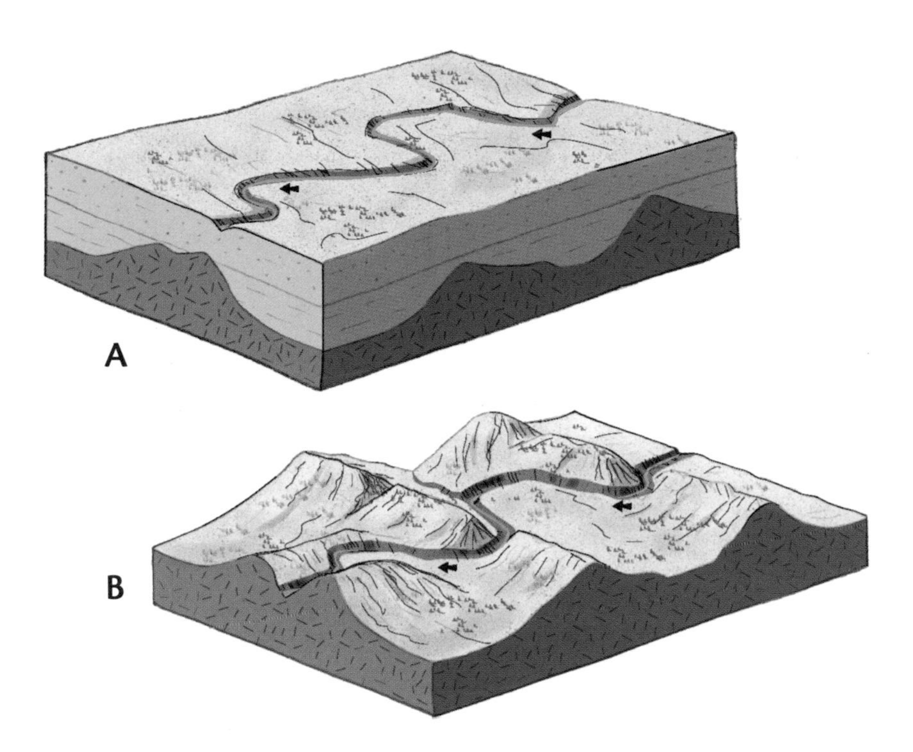

A Superposition begins when a river flows across a subdued landscape that is underlain by relatively soft rock. **B** As this soft covering is stripped away by erosion, a preexisting buried topography is exposed and can be cut into by the river.

that he quoted Marvine word for word in his own 1875 report. Powell claimed to have had the same interpretation for the streams in the Rockies from his fieldwork during the summers of 1867 and 1868. He recognized the importance of this idea and proposed a name for it, too: *superimposed valleys* (today the term is shortened to *superposed*). Superposed valleys originate on higher, subdued surfaces but work their way progressively down into preexisting, buried landforms.

When Powell wrote his 1875 report, however, he insisted that the area of the Green River adjacent to the Uinta Mountains formed by antecedence and not superposition. He seems to have struggled with sorting out the two possibilities, saying that he checked and rechecked the field relations, but ultimately settled firmly in the antecedent camp.

Regarding the Grand Canyon and the Colorado River, Powell actually had little to say about their origin. He did note the apparent lack of association between the many faults in the canyon and the course of the river. With this observation he concurred with Newberry that faulting did not form the canyon. However, he made perhaps too great a leap when he stated:

> *All the facts concerning the relation of the water-ways of the region to the mountains, hills, cañons, and cliffs, lead to the inevitable conclusion that the system of drainage was determined antecedent to the faulting, and folding, and erosion.*

This was a time when the bigger picture of the evolution of the West's landscape was completely unknown. Plate tectonic theory was nearly one hundred years in the future, and Powell did not have the luxury of knowing the age of the uplifts that crossed the path of the river; his role was simply to establish a framework of generalized relationships. More important, he couldn't have given much thought to the Grand Canyon's origin, simply because he was in a race for survival against it (at least on his first river trip). He took what he more carefully observed upstream on the Green River and tried to make it fit with what he observed in the Grand Canyon. His pioneering explorations served to mentor and inspire two successive generations of southwestern geologists, including Grove Karl Gilbert, Clarence Dutton, and Charles D. Walcott.

Superposition

Clarence Dutton, 1882

Courtesy of USGS

Grove Karl Gilbert followed in many of the same footsteps as his mentor, John Strong Newberry, and traveled by boat up into the Grand Canyon as far as Diamond Creek. He ultimately rode overland to the north side of the river, going down into the canyon along Kanab Creek. He has the distinction of being the person who named and identified the Colorado Plateau, the Basin and Range, as well as naming Grand Canyon's first rock formation, the Redwall Limestone. Thus, at the beginning of the 1880s, three geologists had observed the Grand Canyon—Newberry, Powell, and Gilbert.

The fourth geologist to visit the canyon and ponder its development was Clarence Dutton, a Yale-educated army officer who, from viewpoints he named, including Point Sublime and Toroweap Overlook, immediately

Artist William Henry Holmes accompanied Clarence Dutton on his survey of the Grand Staircase and Grand Canyon's Toroweap area and painted the first staggering images of this unique landscape. The people and horses near the water pocket provide scale. *The Grand Cañon at the Foot of the Toroweap–Looking East* by William Henry Holmes, 1882

recognized the geologic significance of the landscape laid out before him. He approached the Grand Canyon after traveling through the high plateaus in southern Utah and was able to make the important evolutionary connection between the two. He recognized that strata composing the Grand Staircase (a name he also conceived) once covered the Grand Canyon region, only to be stripped away in what he called the "Great Denudation." He theorized a later period of canyon cutting, which he termed the "Great Erosion." Dutton, therefore, was the first geologist to differentiate between two cycles of erosion: one that created the Grand Staircase by the lateral stripping away of strata and one that created the Grand Canyon through vertical dissection. These two different episodes of erosion created the landscape we see today.

Clarence Dutton wrote what is perhaps the most readable scientific report ever written, *The Tertiary History of the Grand Cañon District*. Published in 1882 as a monograph of the U.S. Geological Survey, this report summarized his observations and his concurrence with Powell for an antecedent origin of the river.

> *The river is older than the structural features of the country. Since it began to run, mountains and plateaus have risen across its track and those of its tributaries. . . . As these irregularities rose up, the streams turned neither to the right nor to the left but cut their way through in the same old places.*

However, he added something original of his own to Powell's version that addressed the way in which the Colorado had been positioned in its exact course.

> *What then determined the present distribution of the drainage? The answer is that they were determined by the configuration of the old Eocene Lake bottom at the time it was drained.*

This is a notable addition. Here, Dutton proposes a superposed origin (see diagram on page 59) for the river's precise positioning on the landscape. He states that when the Eocene-age (a subdivision of the Cenozoic Era, lasting from 56 to 34 million years ago) lakes dried up, the river flowed across these lake beds. He returns to antecedence later in his report by proposing that the folds and faults became active only after the shape of the lake bottom had already determined the course of the river.

Dutton was one of the greatest minds to ever ponder the origin of the Grand Canyon landscape; however, it was beyond him to formally challenge the prevailing view of Powell regarding antecedence. His greatest gift to us may be his appreciation of the Colorado Plateau landscape, for he was perhaps the first person to view it as intrinsically beautiful or even nurturing. His writing signals a collective change in the view people would have regarding this exhilarating landscape.

> *The lover of nature whose perceptions had been trained in the Alps . . . or . . . the Appalachians . . . would enter this strange region with a shock, and dwell there for a time with a sense of oppression, and perhaps with horror. Whatsoever things he had learned to regard as beautiful and noble he would seldom or never see, and whatsoever he might see would appear to him as anything but beautiful and noble. . . .*

The colors would be the very ones he had learned to shun as tawdry and bizarre. . . .

But time would make a gradual change. Some day he would become conscious that outlines which at first seemed harsh and trivial have grace and meaning; that forms which seemed grotesque are full of dignity . . . that colors which had been esteemed unrefined, immodest, and glaring, are as expressive, tender, changeful, and capacious of effects as any others. Great innovations, whether in art or literature, in science or in nature, seldom take the world by storm. They must be understood before they can be estimated, and must be cultivated before they can be understood.

The Butte Fault

Courtesy of USGS

Charles D. Walcott, 1890

Charles Walcott is best known for his work on the unusual fossils of the Burgess Shale in the Canadian Rockies. However, in 1882 his field party constructed the Nankoweap Trail in Grand Canyon, which gained him access to the faults and folds in eastern Grand Canyon. The Butte Fault, which he named, marks the eastern edge of the Kaibab upwarp, and Walcott determined that this fault and the associated East Kaibab monocline were actively uplifting the Kaibab Plateau before the overlying Mesozoic rocks were stripped away to the northeast. This hinted that the fold could be much older than previously thought. He writes in his report:

It is difficult to understand how the cañon could have existed even to a limited depth, in its present position, at the time of the elevation of the Kaibab Plateau. . . . If followed out in all its bearings, [this] would probably necessitate some change in the now accepted views concerning the manner of erosion . . . of the Grand Cañon.

Walcott did not verbally challenge Powell's theory of antecedence, but his words suggest a certain persistent discomfort with the idea. In his report, he noted how the Colorado River in Marble Canyon exactly parallels the trace of the Butte Fault for ten miles from Nankoweap Creek to the Little Colorado. This showed unequivocally that the river must have

become positioned after that structure was formed. In this way, he became the first geologist to significantly challenge the prevailing idea that the Colorado River was older than the faults and folds that cross its path.

Attack on Antecedence

Courtesy of USGS

Samuel F. Emmons, 1897

If Walcott delivered the first shot at Powell's theory of antecedence for the Colorado River, then Samuel F. Emmons, of the U.S. Geological Survey, fired the cannons. He published a short article in the journal *Science* based on work he completed in the area of the Uinta Mountains and the Green River in 1872. In it he vigorously sought an explanation from Major Powell on how the Green River could be older than the Uinta uplift when eight thousand feet of Eocene lake sediment was at least partially derived from the uplifted mass and the Green River later cut through those same sediments:

> *What then became of the river while these eight thousand feet of Tertiary sediments were being deposited? It could hardly have continued its course at the bottom of the . . . lakes. . . . If it ceased to flow during this time its bed must have been filled with [lake] sediments . . . and when the lakes finally drained, it is hardly conceivable that in re-determining its course . . . it should have attacked the flanks of the Uinta range . . . at exactly the same point it had entered before.*

In this paragraph, Emmons offers a convincing argument about the age of the Green River relative to the Uinta uplift. He points out that the river cut through lake deposits that were derived from the adjacent uplift. This suggested to him (and others who followed) that the lake deposits must be younger than the uplift. Since the river sliced through these deposits, then the river must be younger than the sediments and the uplift. Emmons

Charles D. Walcott studied and named the Butte Fault, which marks the eastern boundary of the Kaibab upwarp. Walcott recognized that the fault must be older than the river, since the river seems to have been emplaced parallel to it. Photograph by Wayne Ranney

writes that he tried to engage Major Powell in an explanation for his line of reasoning, but "he [said] it is so long ago he no longer remembers the course of reasoning he followed at the time." Emmons concludes:

> *Whatever may be the outcome of such an examination, it would seem proper that the antecedence origin of this river should be held in abeyance until some positive evidence of it can be furnished.*

---

he nineteenth century closed with a few preliminary attacks on the theory of antecedence, yet no alternative was poised to take its place. A basic framework of ideas had been adopted, but an increasing recognition of the relationship between the river and the landscape led to a change in thinking that would shake the foundations of antecedent thought.

EARLY TWENTIETH CENTURY

The twentieth century began with geologists undertaking field excursions to Grand Canyon in horse and buggy and closed with them circling the earth in space and taking photos of the canyon from above. This degree of advancement in technology is astounding and might suggest a similar advance in our understanding of the canyon's origin. However, many of the same dilemmas that confronted the horse-and-buggy-bound geologists still haunt us today: How old is the river? What specific processes carved the canyon? What is the timing and frequency of plateau uplift?

The Plateau and Canyon Cycles of Erosion

Harvard University Archives, HUP Davis, William Morris (4)

William Morris Davis, 1901

In June 1900, William Morris Davis, known as the father of geomorphology (the study of landforms), completed a twenty-three-day overland trip from Flagstaff, Arizona, to Milford, Utah. He traveled by horse and wagon, averaging twenty-five miles a day. His route took him north along the Echo Cliffs to Lees Ferry, where he crossed the river and continued beneath the Vermilion Cliffs to the Kaibab Plateau and Fredonia, then proceeded to the Toroweap Valley and a stunning view of the Colorado River below (see map on pages 18–19). Davis was an astute

Vermilion Cliffs, seen on the horizon from this North Rim vantage, are part of the Grand Staircase. Photograph by Chuck Lawsen

observer of landscapes and took a keen interest in how the Grand Canyon may have formed.

Davis used the foundation that his nineteenth-century colleagues had laid for him and suggested ways that their findings could be refined. He affirmed Dutton's idea that two cycles of erosion were responsible for the present arrangement of deep canyons set upon the broad plateaus. The first cycle, occurring during the Paleogene Period (65 to 23 million years ago), involved the lateral stripping of Mesozoic rocks from on top of the Grand Canyon area, and he termed this the "plateau cycle," correlating it with Dutton's Great Denudation. He called the second cycle, occurring in the Neogene Period (23 million years ago to the present), the "canyon cycle," relating it to Dutton's Great Erosion, which involved the relatively recent cutting of the canyons. He argued broadly for a humid climate during the plateau cycle and an arid climate for the canyon cycle. His perceived relationship of erosion style to climatic factors remains relevant today.

Davis noted the relationship between the river and the Butte Fault in eastern Grand Canyon (as Walcott had done before him) and indicated that the Little Colorado and Marble Canyon reaches of the river may have been positioned in a subsequent valley between the Kaibab upwarp to the west and the now-eroded Jurassic sandstones to the east (ancestors to the present-day Echo and Vermilion Cliffs). These observations show that he was not completely comfortable with an antecedent origin for the river, arguing that the river was likely positioned in response to the preexisting faults, folds, or topography. He writes:

> *It is not my intention to discount such [theories] . . . but only to emphasize the opinion that the facts now on record, combined with such knowledge of the region as our party was able to gather . . . warrant the consideration of at least one hypothesis alternative to the theory of antecedence, as an explanation for the origin of the drainage lines in the Grand Canyon district. I do not on the one hand consider the antecedent origin of the Colorado disproved, but, on the other hand, such an origin does not seem compulsory. The chief objection to the theory of antecedence is not that rivers cannot saw their way through rising mountains . . . but rather that this theory makes a single stride from the beginning to the end of a long and complicated series of movements and erosions, overlooking all the opportunities for drainage modifications on the way.*

Davis's words reflect respectful diplomacy toward Powell while at the same time challenge his theory. According to Davis, drainage in the Grand Canyon area was directed initially to the northeast down the steps of the east-dipping monoclines and toward the Grand Staircase. Later, as the Basin and Range was formed and blocks of the earth's crust were lowered down to the west, this drainage became reversed both by Basin and Range faulting and reversed tilting of the entire plateau. The deep dissection of the canyon only occurred after faulting and the reversal of the drainage. These were bold concepts that reinvigorated thinking on the origin of both the river and the canyon, and many of these ideas may still ring true today.

Davis also offered cogent observations into the possible origin of the Esplanade Platform, the broad terrace beneath the Hermit Formation in western Grand Canyon, saying that it was not the result of the river meandering freely across that surface for an extended period of time (and before incision of the inner canyon). Instead, he pointed out that everywhere the Esplanade is found within Grand Canyon, it is generally equidistant from either the North or South Rim and the river. He postulated that the inherent softness of the Hermit Formation upon exposure to erosion caused the upper cliffs on both sides of the canyon to retreat at an equal rate away from the incising river. His interpretation is likely correct.

All of these observations prompted Davis to state that even though the Grand Canyon was extremely deep and rugged, the Colorado River appeared to be eroding it at a slower rate than in the past. Davis came to this conclusion after noting the relatively low gradient of the river in the canyon (averaging only about eight feet per mile) and observing that almost all the tributary streams in the canyon enter the river at grade (the same level), even though many of them are dry most of the year. He claimed that the level of the streams might be the result of erosion rates having been greater at some earlier time; otherwise, how could a stream that is dry most of the time erode as deeply as the Colorado River? His recognition of this is a phenomenal insight, and although we can now explain tributaries entering the Colorado at grade by invoking other processes, his observations allow us to consider how streams, even dry ones, deepen their canyons.

All of these insights were fresh and original, and stimulated others to look critically at the evidence for how long the river might have been at work in carving the Grand Canyon. Davis offered a synthesis of what he observed on this trip and a second field trip conducted in 1902:

> *The most emphatic lesson that the canyon teaches, is that it is not a very old feature of the earth's surface, but a very modern one; that it does not mark the accomplishment of a great task of earth sculpture, but only the beginning of such a task; and that in spite of its great dimension, it is properly described as a young valley.*

In the first decade of the twentieth century, other geologists contributed ideas that addressed the origin of the Colorado River. Willis Lee (1906) proposed that the river once flowed through the Sacramento and Detrital Valleys, located between Kingman and Las Vegas. He used the presence of gravels in those valleys as evidence (since found to be merely washed from the surrounding mountains). Harold Robinson (1909) documented the existence of a broad erosion surface that formed on and adjacent to the Colorado Plateau before the canyon cycle of erosion. According to him, a part of this old surface is preserved under the lava flow on Red Butte, south of Grand Canyon. Douglas Wilson Johnson, on a wagon trip in 1906 from Prescott, Arizona, to Salt Lake City, Utah, verified the findings of Walcott and Davis that the faults and folds of the region seem to be older than the river. As more eyes looked upon the canyon landscape, an idea began to emerge that the Colorado River might have attained its present configuration much later than initially envisioned. The time had come for a newer theory of the river's origin, one that would speak to the evidence that suggested the river might be quite a bit younger than previously thought.

Basin Spillover Theory

Courtesy of USGS

Eliot Blackwelder, 1934

Most geologists prior to the 1920s had argued that the Colorado River was old—at least Paleogene in age (or more than about 50 or 40 million years old)—but Eliot Blackwelder would offer a challenge to this prevailing view. He sought to dismiss Lee's postulation that the river once flowed south through the Sacramento and Detrital Valleys by showing that the gravels located there were of local derivation only and not deposited by the Colorado. His view encompassed the entire river system from its headwaters to its mouth and noted the curious setting of the river flowing alternatively through open valleys and narrow, confined canyons. With these observations, he sought to advance the idea that the Colorado might actually be a young river.

Blackwelder observed the presence of many closed basins adjacent to, but not part of, the river system south of Las Vegas. He questioned why, if the river was old, had it not already extended tributaries up into them, thus capturing their drainage? He looked at other southwestern rivers such as the Owens, Amaragosa, and Sevier, and noticed how they had developed their courses by sequentially filling basins along their path with water that successively overflowed into the next basin, creating a string of lakes that became connected with steep-gradient rivers confined in narrow canyons. Blackwelder envisioned such a pre–Colorado River landscape in the arid region of the Grand Canyon. Later uplift of the Rocky Mountains would create a catchment for moisture that would flow south into these interior basins:

> *The present existence of the Colorado River is due solely to the fact that the Rocky Mountains in Colorado, Wyoming, and Utah are sufficiently high to condense moisture. . . . It is reasonable to infer that, as the region bulged upward, the local streams on the higher and more northerly mountains extended themselves [southward], forming lakes in the nearest desert basins. As this influx exceeded evaporation . . . the lakes rose until they overflowed the lowest points of their rims and spilled into adjacent basins. In time, enough excess outflow may have developed to fill a series of basins all the way to the Gulf of California, thus forming a chain of lakes strung upon a river.*

Blackwelder knew that such an assertion was highly speculative, but he used what he saw in the present-day landscape to formulate a viable story for a young Colorado River. Despite perhaps being too much of a stretch for many geologists at that time to consider such a relatively simple hypothesis for the large and complex Colorado River system, he presented a credible alternative to the previous theories that stated the Colorado River must be older than the uplifts that enclosed it. Interestingly, his idea has experienced a revival among modern geologists, especially in relation to the lower Colorado River from Hoover Dam to the Gulf of California. Blackwelder's closing argument is perhaps the most cogent summary given up to this point in time:

> *The foregoing sketch of the origin and history of the Colorado River is frankly theoretical. Science advances not only by the discovery of facts but also by the proposal and consideration of hypotheses, provided always that they are not disguised as facts. This view will not meet with general acceptance. There are doubtless many facts unknown to me that will be brought forward in opposition. Perhaps their impact will prove fatal to the hypothesis. In any event, the situation will be more wholesome, now that we have two notably different explanations, than*

it was when it was assumed by all that the river had existed continuously since [Paleogene] time. It seems to me that the new hypothesis is harmonious with most of the important facts now known about the geology and history not only of the Colorado River but of the Western States in general.

To this day, Eliot Blackwelder remains one of the giants in the evolution of ideas regarding the origin of the Colorado River. His challenge to previously accepted views opened new avenues of thought regarding how and when the river may have formed. As we shall see, some prominent modern geologists accept the idea that basin spillover offers the best explanation for how some portions of the Colorado River system was integrated.

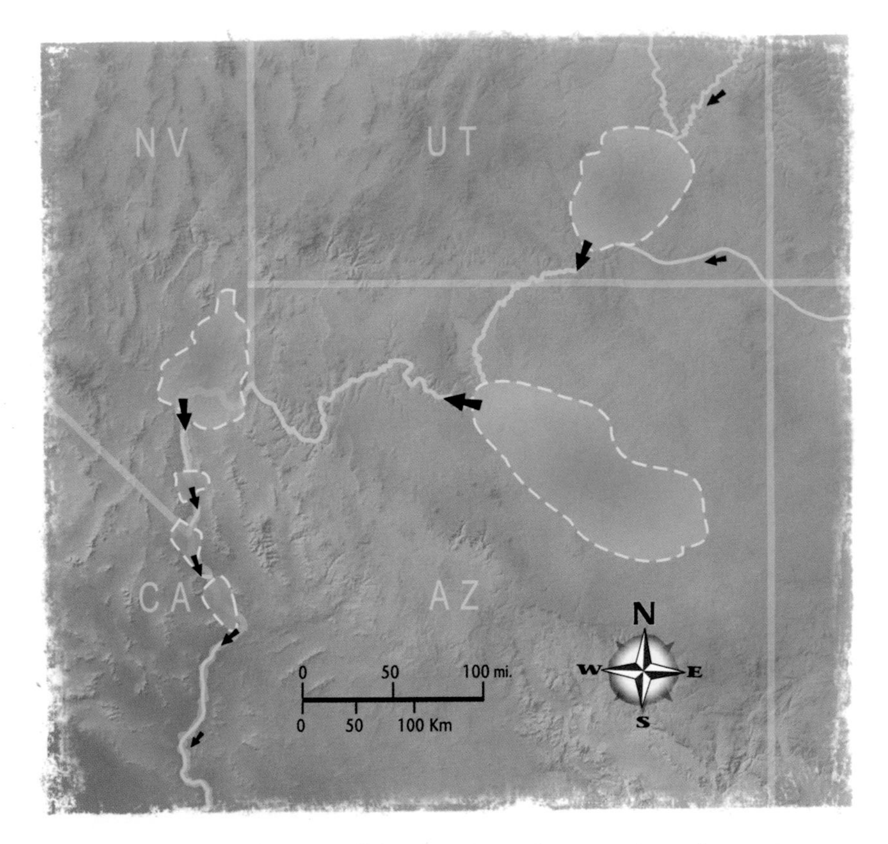

Blackwelder envisioned a string of lakes from the Rocky Mountains to the sea, along which the course of the Colorado River was formed through spillover.

The Racetrack Theory

Donald Babenroth and Arthur Strahler, 1945

Up to this point, few scientists had specifically addressed the problem of why the Colorado River abruptly turns ninety degrees into the barrier-like Kaibab upwarp in eastern Grand Canyon near Desert View. Geologists had long recognized this confusing relationship between the river and the upwarp, but no one had tackled the problem head on. That would change in 1945 when Donald Babenroth and Arthur Strahler published a paper that stated:

> *The feature requiring special attention is that the river passes from the relatively low Marble Platform area westward through the Kaibab arch, which resembles a giant anticlinal barrier in the path of the river.*

The authors first reviewed the ideas of Powell, Dutton, Walcott, and Davis and discarded Davis's idea of backtilting of the plateau and river channel, stating that the amount of reversal necessary was too much

The East Kaibab monocline is a flexure in the earth's strata that elevates the Kaibab upwarp and North Rim (upper left) three thousand feet higher than the Marble Platform (right).Photograph by Michael Collier

given the evidence at hand. They offered a specific explanation for why the Colorado River appears to cross the high-standing, barrier-like Kaibab upwarp as it turns westward beneath Desert View. They noted four possibilities to explain this odd arrangement: antecedence, superposition, consequence, or subsequence (see sidebar on page 46). In the end they favored a subsequent origin for the Colorado River in this area (granting that the other three might be possible for other parts of the river):

> *In general plan, the Colorado River makes a great bend [westward] around the pitching nose of the Kaibab arch. This suggests that, as the Mesozoic strata were being stripped from the region, the river may have occupied a subsequent lowland belt of weak Moenkopi and Chinle shales between the plunging nose of the Kaibab limestone arch and the encircling north-facing cliffs of Jurassic sandstones. . . . The subsequent valley would have been several miles wide and would have coincided approximately with the present Grand Canyon.*

Babenroth and Strahler offered a logical explanation for why the Colorado River only *appears* to turn west into the uplifted Kaibab upwarp. They envisioned that the river once flowed at the seven-thousand-foot elevation and was confined in a low valley that was

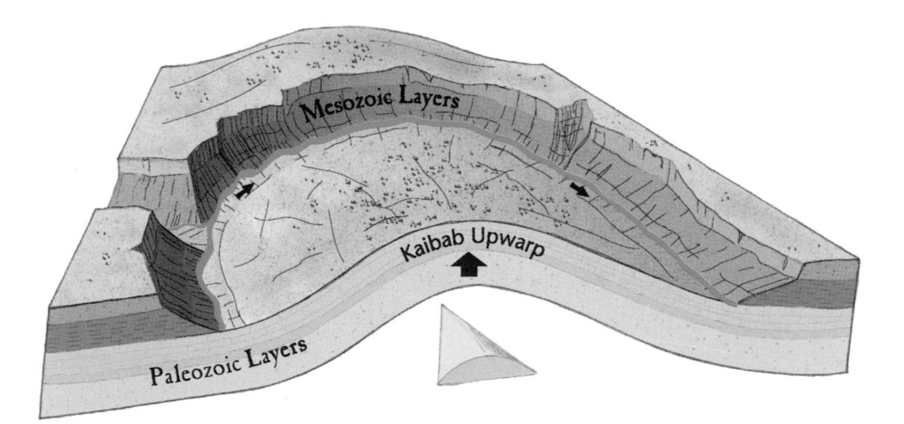

Babenroth and Strahler suggested how the Colorado River may have been positioned in its course near the Great Bend at Desert View. They thought that a river once flowed through a subsequent valley located between a receding cliff of Mesozoic rock on the south and the arch of the Kaibab upwarp on the north.

bordered between the south-facing slope of the Kaibab upwarp (essentially the surface of the North Rim today) and Mesozoic rocks that were being stripped off this surface toward the south (essentially where the South Rim is located, with Red Butte and Cedar Mountain as the only visible remnants of this once more extensive cliff). Their theory necessitated that the Marble Platform was buried in a thick succession of Mesozoic rocks such that the Kaibab upwarp was topographically indistinct from the land that surrounded it to the east. How else could the river flow toward the west, from the area of the Marble Platform across the Kaibab upwarp?

Babenroth and Strahler were among the first to attempt to explain how the Colorado River crossed the Kaibab upwarp. Later geologists would advance and popularize their idea, adding the moniker the "racetrack theory." The name comes from the way the receding cliff of Mesozoic strata on the south side of the river resembles the way a grandstand (the cliffs) swerves around the course of a racetrack (the river).

The Muddy Creek Problem

Courtesy of USGS

Chester Longwell, 1946

Before and during the construction of Hoover Dam, Chester Longwell studied deposits near the Colorado River that would ultimately be submerged in the rising waters of Lake Mead. How ironic that with the construction of the dam and the subsequent taming of the free-flowing Colorado River, important insights into the river's wild and ancient origins would be discovered. A professor of geology at Yale University from 1920 to 1956, Chester Longwell was fascinated by the emerging dichotomy regarding the age of the Colorado River:

> *One of the major unsolved problems of the region is the date of origin of the river itself. . . . Geologists who have no direct acquaintance with the region will be at a loss to understand so wide a divergence in interpretation.*

How true these words are even today. In the paper "How Old Is the Colorado River?," Longwell summarized the two extreme views: Eliot

The western terminus of the Grand Canyon is located where the Colorado River exits the Grand Wash Cliffs. The area is now flooded by water (or sediment) from Lake Mead. Partially eroded remnants of the Muddy Creek Formation can be seen in the foreground and in the low hills across the reservoir. Photograph by Wayne Ranney

Blackwelder's theory that the river developed from a string of overtopped lakes in only the last few million years and the evidence invoked by other geologists who thought the river was much older. Longwell looked at various deposits near the base of the Grand Wash Cliffs and concluded that the vast majority of them were not derived from the Colorado River. These deposits were later named the Muddy Creek Formation and were found to be as young as 6 million years. Longwell's conclusion was that the river, and thus the Grand Canyon, could be no older than the deposits. Because deposits for an older Colorado River were found far upstream of Grand Canyon, the evidence here of a quite young river was puzzling, and this relationship later was referred to as "the Muddy Creek problem." Longwell elaborated on the importance of the evidence:

There is no possibility that the river was in its present position west of the Plateau during Muddy Creek time. The suggestion occurs that a stream, either permanent or intermittent, may have developed on the site of the present Grand Canyon, and debouched [emerged] into closed basins west of the Plateau. However, if such a stream had any considerable length, it should have contributed rounded pebbles representing the varied lithology east of the Grand Wash Cliffs. No such stream-worn pebbles have been found in the basin deposits. . . . Deposits made by the through-going Colorado are unmistakable. They contain, as a characteristic ingredient, gravel made up of rounded pebbles and cobbles, representing a large number of bedrock types found in the Plateau.

Since the middle of the twentieth century, this critical piece of evidence has served as the pivot point around which all other ideas regarding the age of the Colorado River have revolved. As we shall soon see, later geologists would be required to devise clever and sophisticated ways for the Colorado River to flow away from, around, or even beneath the Grand Wash Cliffs, at the base of which the Muddy Creek problem was first recognized. Even today, the Muddy Creek problem remains the nexus through which many other theories must go.

Longwell also noted a very important piece of information regarding the relationship of the Colorado River's development to the larger development of the western landscape:

Attempts to project the present drainage system as far back as early [Paleogene] encounter possible difficulty in some significant facts of the regional history. Intensive orogeny [mountain building], with development of great thrusts and folds . . . affected a wide belt directly west of the Plateau near the close of the Cretaceous Period [145 to 65 million years ago]. Presumably therefore, a mountain zone extended across the course of the present Colorado River in early [Paleogene] time, with . . . topography opposed to any westward drainage from the lower Plateau area.

What Longwell is referring to is the existence of the Mogollon and Sevier Highlands, ancient mountain belts located south and west of the Colorado Plateau until the Basin and Range Disturbance destroyed them beginning about 17 million years ago (the Basin and Range Disturbance is the tectonic event that formed the Basin and Range Province). Longwell pointed out that an early incarnation of the Colorado River could not conceivably flow toward uplifted mountains. The crucial point is that any precursor to the Colorado River could not have developed before the Basin and Range was lowered by faulting, beginning 17 million years ago. More probable, he stated, was that drainage was directed to the northeast onto the southern Colorado Plateau and that it possibly pooled there in enclosed basins. Longwell astutely reminded geologists that they seemed to have ignored this salient fact:

> *In outlining the foregoing hypothesis, it has been assumed that the Plateau has had exterior drainage continuously though the [Cenozoic Era]. However . . . the region probably was unable to support a through flowing stream like the Colorado for a considerable period. . . . During such an interval the drainage of the Plateau area would have been accomplished by intermittent streams ending in a number of separate closed depressions, as in the Great Basin at present. . . . When the [Rocky Mountains] attained such altitude that increased precipitation, [they] supplied a surplus of runoff into the Plateau, the configuration of the surface may have been such as to guide the overflow along a new consequent course to the west.*

In this passage, Longwell appears to be lining up behind the hypotheses of Eliot Blackwelder by favoring the idea that before the modern Colorado River was born, the plateau surface may have been the site of closed interior basins. The crux of the matter here is summed up by some modern geologists when they ask, Where are the supposed basin deposits? To which supporters reply that perhaps there never were any or that they could have been removed during the most recent phases of plateau uplift and erosion. In light of Longwell's observations, one sees that controversies and opposing viewpoints were beginning to form regarding the origin of the Colorado River. It is at this point that we see theories beginning to converge and clash with one another.

Stream Capture by Headward Erosion

Courtesy of USGS

Herbert Gregory, 1947

Starting at the turn of the twentieth century, several geologists cited evidence for the existence of plateau-wide erosion surfaces that formed during the period of lateral stripping, also known as Dutton's Great Denudation and Davis's plateau cycle. Herbert Gregory had accompanied Davis on his 1900 expedition, and influences from Davis had caused him to become an erosion-surface advocate. However, Gregory later changed his mind after many field seasons scampering across the plateau, and by 1947 he considered that the erosion surfaces were not plateau-wide features but rather only localized structures (pediments), or that the erosion surfaces formed simply when soft strata was eroded off the top of resistant beds. The distinction between a truly beveled surface (that cuts across formations and is not confined to a bedding plane) and one formed solely from stratigraphic resistance is quite subtle and not easily discerned.

Gregory's contribution to understanding the history of the Colorado River concerned the uplift of the high plateaus in Utah providing the elevation necessary for increased precipitation:

> *Guided by topography, rains and melting snows doubtless spread a net of short . . . streams over the plateau tops. . . . It may be conjectured that the south-flowing waters from the uplifted, warped, and fractured highland found a resting place on the structural lowlands in the approximate position of the present Grand Canyon and the . . . Little Colorado Valley; that their most easily developed outlet to the sea was westward; and that along this course the new-born stream, locally expanded into lakes, descended from one to another plateau block over high cliffs, and thus developed energy rarely acquired by running water.*

Gregory's idea that the runoff found a resting place in the region of the Grand Canyon and the Little Colorado Valley is an interesting one, for he continues:

> *It seems probable that the original Colorado was a relatively short stream of moderate volume fed from the north by the ancestral Virgin,*

Legends and Other Creation Myths

Long before scientists studied the Grand Canyon, native peoples shared legends about how the canyon and the river were created. These legends are not scientific theories, but they do coincidentally mimic some modern scientific ideas. These legends are not meant to give a chronology of natural events but rather serve to affirm some culturally significant idea. If any of us had been alive more than 150 years ago, these legends would absorb our thoughts as we listened to our elders recall the creation days.

Five modern tribes live in or adjacent to the Grand Canyon—the Havasupai, Hualapai, Southern Paiute, Navajo, and Hopi—and several other tribes maintain close ties to the canyon. Earlier cultures preceded these tribes, and archaeological evidence attests to their former presence, although what their thoughts might have been regarding the Grand Canyon's origin is unknown to us. The earliest evidence of human occupation in the Grand Canyon dates back more than four thousand years, but what these split-twig figurine makers thought about the canyon is lost in time. Perhaps even the ice age hunters were here almost twelve thousand years ago; A fragment of Clovis point (about 12,000 BP) and a waste fragment from a Folsom spear point (about 9,000 BC) have been found at Grand Canyon.

American Indian creation stories relating to the Grand Canyon typically invoke gods with supernatural powers that create catastrophic floods or similar events to create the river and canyon. Floods are a theme common to the stories of many agricultural groups who live along rivers or shorelines throughout the world. An anthropologist who once studied the Havasupai Indians in Grand Canyon recorded this creation story: "One day a mischievous god, named Ho-ko-ma-ta, started a rainstorm greater than a thousand Hacka-tai-as (Colorado Rivers). The benign god, Toc-ho-pa, wanted to save his daughter, Pu-keh-eh, from the ensuing flood and put her in a hollow piñon pine log. As the floodwaters rushed to the sea, they created Chic-a-mimi (Grand Canyon) and Pu-keh-eh was able to crawl out of her log safely and become the progenitor of all human beings." This story astutely associates the cutting of the canyon with the Colorado River; however, the legend deviates from the scientific evidence when it describes the whole process as happening during a single flood. No geologist believes that the canyon formed in this way.

Ancestral Puebloan petroglyphs can be found in a dry wash along the Colorado River near Tanner Rapid. Photograph by Larry Lindahl

John Wesley Powell recorded a Paiute legend regarding the origin of Grand Canyon: Long ago, there was a great and wise chief who mourned the death of his wife and would not be comforted until one of the Indian gods came to him and told him she was in a happier land. The god offered to take him there that he might see for himself if, upon his return, he ceased to mourn. The great chief promised. Then the god made a trail through the mountains ... This trail was the cañon gorge of the Colorado. Through it he led him and, when they had returned, the deity exacted from the chief a promise that he would tell no one of the joys of that land lest, through discontent with the circumstances of this world, they should decide to go to heaven. Then he rolled a river into the gorge, a mad, raging stream that should engulf any that might attempt to enter thereby.

In this instance the canyon was created independently of the river.

A book published in 2003 documented a Western religious interpretation for how the canyon was formed. Its premise is that all of Grand Canyon's flat-lying strata, along with those still seen at Zion and Bryce Canyon, were deposited during Noah's flood. Recession of this water completely removed the Zion and Bryce Canyon rocks from the Grand Canyon area and carved the great gorge at the same time. According to this story, all of these events—the deposition of fifteen thousand feet of sediment, the complete removal of the upper half of the deposits, and the cutting of the Grand Canyon—happened during a single flood in *one year*. Geologists use evidence from many different sources to show how deposition of strata occurred over a much longer time span with a completely separate set of events to carve the canyon at a much later time. The idea that all of this geology could have occurred in one year is inconceivable to anyone who has critically and thoughtfully looked at the evidence.

More troublesome to scientists, however, is the manner in which this "research" was conducted. True science does not proceed by drawing conclusions first (there was a flood that covered the whole earth) and then identifying those isolated pieces of the evidence that make a conclusion seem sound (the Grand Canyon must be a result of this flood). A scientist looks at all the possible ways a landscape could have evolved and then tests his or her hypothesis to determine if it is sound. In this way, some theories will survive the test of rigorous examination while others will not. Creationist ideas are based on religious faith and the belief that the Bible contains a record of all of earth history. This is troublesome, even to many people of faith. Contrary to what creationists may say, this faith-based approach is not science. As with any idea, concepts should be approached with an open mind, studied carefully, and critically considered.

Kanab, and Paria rivers, and from the south by the Little Colorado, which gathered water from the large region between Navajo Mountain and the Mogollons.

Gregory theorized that the "young" Colorado was able to capture drainage area north of the Grand Canyon region from other nearby rivers because of its steeper gradient. This is how it became the integrated river we find today. Thus, in the debate about the Colorado's origin, he may have been the first geologist to propose the idea of what would later be called stream capture by headward erosion.

The first half of the twentieth century ends with geologists increasingly arguing for a young Colorado River. These ideas contrast with the earliest views that the river was old. In the next fifty years, geologists would attempt to resolve these seemingly conflicting views.

LATE TWENTIETH CENTURY

During the second half of the twentieth century, field studies related to the Grand Canyon's origin increased tremendously. The search for uranium, prompted by the beginning of the Cold War, initiated numerous field studies on the plateau, and some geologists became tangentially interested in the origin of this captivating landscape. One of the watershed events during this half century was a symposium held at the Museum of Northern Arizona in 1964 aimed specifically at the question of how the Colorado River in Arizona had evolved.

Anteposition

Courtesy of USGS

Charles Hunt, 1956

One of the great names in the field of Colorado River studies is Charles (Charlie) Hunt of the U.S. Geological Survey. His classic paper "Cenozoic Geology of the Colorado Plateau" was a synthesis of known information about the plateau, the river, and the landscape. Hunt proposed a possible origin of the Colorado River by constructing ten paleogeographic maps of the Colorado Plateau that showed how the river might have attained its course through time. Hunt offered some intriguing conjectures on the origin of the Grand Canyon as well.

Like many of his predecessors, Hunt recognized the overwhelming evidence for an initial northeast flow of drainage across the southern Colorado Plateau. His maps indicate at least three different freshwater lakes where these streams may have terminated. He reminded readers that long periods of time in which no deposits are preserved make all interpretations necessarily conjectural. Although he accepted that short local drainages may have appeared at the west edge of the plateau by about 30 to 20 million years ago, he thought water was probably still ponded in interior basins on the central plateau at this time. He further suggested that the plateau was uplifted en masse and that superposed tributaries like the San Juan then became entrenched in canyons. He supposed that the rivers would continually adjust their courses as the laccoliths (subsurface intrusions of magma) formed landforms such as the Henry, Abajo, La Sal, and Sleeping Ute Mountains on the plateau.

According to Hunt, as the plateau became higher than the Basin and Range, drainage had to develop somewhere off of the plateau edge. But where? He accepted Longwell's contention that it couldn't have been at the Grand Wash Cliffs because of the Muddy Creek problem. So he postulated that the Colorado River in Grand Canyon might have flowed south through Peach Springs Wash (see map on page 18–19). He was understandably quite cautious about this idea.

Hunt was then left with two possibilities for the origin of the river in Grand Canyon, superposition or stream capture, and he didn't like either of them. Superposition demanded that lake sediments be present as high as the top of the Kaibab Plateau, presently at more than nine thousand feet, and then rise even higher upstream along the river so that it could flow to the southwest. This was untenable to him. Stream capture was also problematic:

> *It would indeed have been a unique and precocious gully that cut headward more than 100 miles across the Grand Canyon section to capture streams east of the Kaibab upwarp.*

Ultimately, Charlie Hunt had to invent and name a process in order to get around the dilemma of the Muddy Creek problem. He called this process anteposition, which incorporates certain aspects of antecedence and superposition. Anteposition proposes that the current path of the Colorado River through the Grand Canyon could have been initially established before the Muddy Creek Formation accumulated (the antecedence aspect). Uplift of the plateau then tilted the river's channel toward the northeast, disrupting and halting its flow into the Muddy Creek basin. This was Hunt's solution to the Muddy Creek problem. He then asserted that the river became ponded north and east of Grand Canyon in the Glen Canyon region and the Bidahochi basin. As these lakes filled, they overflowed to the south and west, initiating deposition of the Hualapai Limestone. Eventually, the Colorado River reestablished its old course by superposition on the lake sediments to the east of Grand Canyon, and it made its way through the Grand Wash Cliffs, dissecting the Hualapai Limestone.

Hunt's ideas were much debated because they were presented at a time when few other original ideas were forthcoming. In the end, however, it seems that he had but few supporters. If nothing else, the theory shows to what extremes geologists will go to explain an enigmatic river whose history is essentially destroyed.

Integration of the River System

Symposium on the Cenozoic Geology of the Colorado Plateau in Arizona, 1967

No treatment of the origin of the Grand Canyon is complete without a discussion of the innovative ideas developed at the Symposium on the Cenozoic Geology of the Colorado Plateau held at the Museum of Northern Arizona (MNA) in Flagstaff, Arizona, in August 1964. For the first time, geologists gathered in one location for the sole purpose of discussing how the Colorado River may have evolved and, perhaps more important, what problems remained to be resolved. The symposium was chaired by Dr. Edwin "Eddie" McKee, a prominent Grand Canyon geologist who through his long, distinguished career worked alternatively for Grand Canyon National Park, the MNA, the University of Arizona, and the U.S. Geological Survey. Twenty geologists attended the ten-day conference and reported on their work from sixteen geographic areas in northern and central Arizona. This collaboration, in which theories and ideas

MNA Bulletin #44, 1967. Courtesy of the Museum of Northern Arizona

were shared and proposals for further research were advanced, resulted in two significant milestones: the development of a timeline that outlined a plausible sequence of events, and an original and provocative theory regarding how the Colorado River and Grand Canyon may have formed from the integration of two separate and distinct river systems. These ideas were published in 1967 as MNA Bulletin #44, "Evolution of the Colorado River in Arizona."

In this seminal report, five stages in the development of the Grand Canyon landscape were proposed:

1. Original northeast drainage on a subdued but sloped surface toward a retreating seaway
2. Modification of this drainage pattern around the rising upwarps of the Colorado Plateau, with flow into freshwater lakes in Utah, Wyoming, and Colorado
3. Renewed uplift of the plateau and the development of three separate drainage systems
4. The existence of interior drainage both east and west of the Grand Canyon (Bidahochi and Muddy Creek Basins, respectively)
5. Development of the modern drainage by the integration of the three river systems, facilitated by continued uplift, headward erosion, and stream capture

Stage 1 refers to the blank canvas of the northern Arizona landscape after the sea retreated for the last time some 80 million years ago. Left in the sea's wake was a surface that gently sloped down to the northeast. Thus, initial drainage in the area originated in the Mogollon Highlands to the southwest and was placed in a direction opposite to that of the modern Colorado River. This landscape scenario is one of the few aspects of the Colorado River's story that is not controversial among geologists.

Stage 2 represents minor modification on this landscape with the differential uplift of structures like the Kaibab and Monument upwarps and the Rocky Mountains. During this stage, drainage flowed from the Mogollon Highlands toward the northeast and into freshwater lakes that formed in the basins between the various upwarps. Lake deposits from this stage are common on the northern Colorado Plateau but remain only as far south as the Bryce Canyon area in southern Utah.

Stage 3 involves the development of three separate drainages in which the Kaibab upwarp acts as a major drainage divide. The western system was called the Hualapai drainage, since its location was near the plateau of the same name. A middle drainage system was proposed and called the Ancestral Cataract-Havasu-Kanab Creek drainage, and went north into Utah (not shown on the diagram on page 87). The final drainage they called the ancestral upper Colorado River, which flowed out of Utah toward the eastern side of the Kaibab upwarp but turned southeast along the present course of the Little Colorado River toward the Rio Grande. This drainage was proposed in part because a system like this was needed to erode suspected lake deposits from southeast Utah.

Stage 4 presupposes that southeast flow of the ancestral upper Colorado was disrupted when uplift near the Arizona–New Mexico state line reversed flow toward the west. This formed a lake near present-day Winslow and Holbrook, and this lake deposited the Bidahochi Formation. At the same time (and taking a cue from Charlie Hunt), the authors proposed that the Hualapai drainage might have flowed south in Peach Springs Wash. Some drainage may have gone west into the Muddy Creek basin, but it didn't leave any sediment with rock types from the Grand Canyon. The authors stated that headward erosion caused by the Hualapai drainage and cutting back into the west side of the Kaibab Plateau could have already begun by this time.

Stage 5 had the Hualapai drainage lengthen its channel to the east, capturing the ancestral upper Colorado River and creating a through-going Colorado River. The capture point was marked at the confluence of the Colorado and Little Colorado Rivers, east of the Kaibab upwarp. The time of this capture event was placed between 10.6 and 2.6

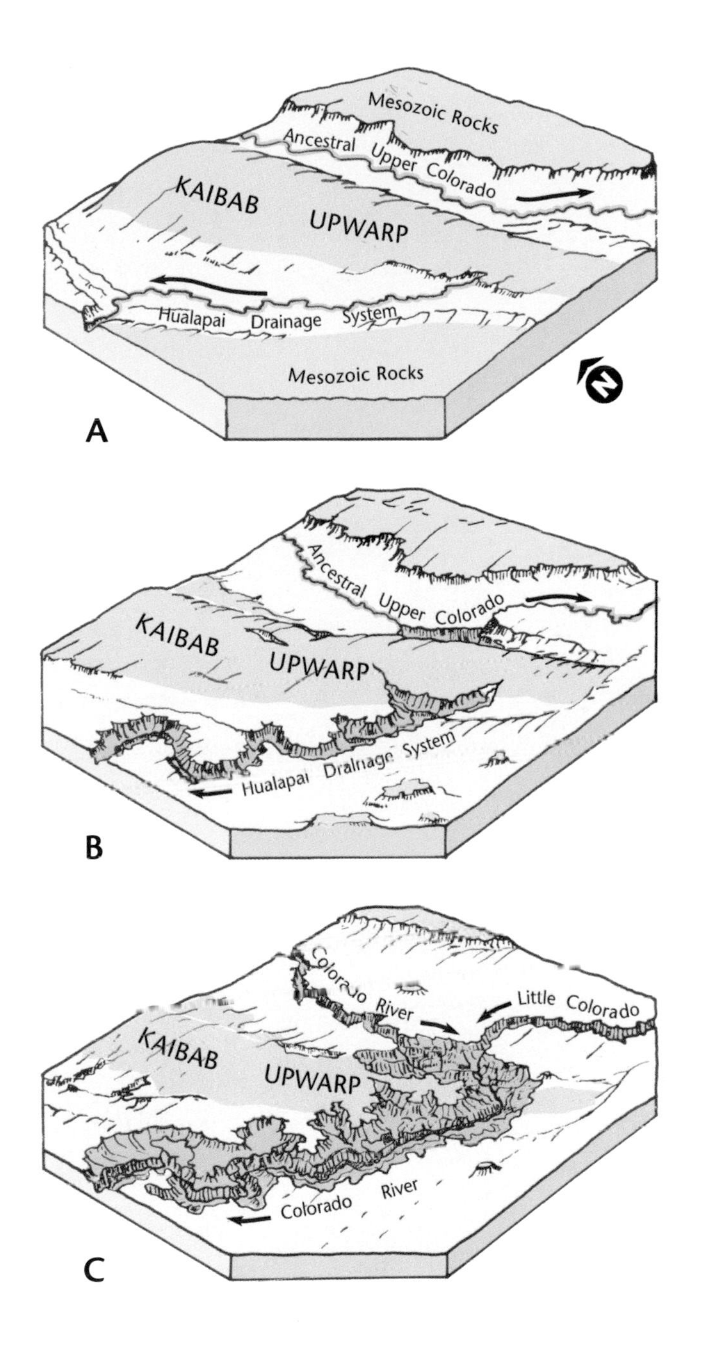

A diagram based on MNA Bulletin #44 shows **A** that the Kaibab upwarp separated the ancestral upper Colorado River from the Hualapai drainage; **B** that headward erosion of the Hualapai drainage facilitated stream capture of the ancestral upper Colorado River; and **C** the subsequent deepening by the integrated system.

million years ago, but advances in dating techniques have greatly refined these ages to between 5.8 and 4.4 million years ago. The newer dates are derived from an ash bed near the top of the Hualapai Limestone (5.8 million years) and a lava flow near Sandy Point on upper Lake Mead that covers unequivocal Colorado River deposits (4.4 million years).

The wider acceptance of the idea that the Colorado River could have been cobbled together from the prior existence of two separate and distinct river systems was a major advance, not only in the understanding of the origin of the Grand Canyon but for the science of landscape evolution as a whole. In our modern world, it may often seem that what we accept as viable theories have always been with us. Yet ideas must be conceived and hatched for the first time by someone who receives a spark of insight for an original idea.

Although some major objections were raised within a few years of the MNA publication (especially about a Rio Grande connection for the ancestral upper Colorado River), the results from this symposium mark a divide in the understanding of the evolution of the river. From this point forward, geologists were not limited to antecedent, superposed, or subsequent origin theories. They could now view the Colorado as an evolving entity that need not be fixed through time. The floodgates were opened to many new and innovative ideas.

Glen Canyon Lake

Charles Hunt, 1969

Courtesy of USGS

In 1969 the U.S. Geological Survey commemorated the one hundredth anniversary of John Wesley Powell's first river trip through the Grand Canyon and published Professional Paper 669. Charlie Hunt authored the paper's major article, "The Geological History of the Colorado River." This monumental work examined the history of the river from its headwaters to its mouth. Hunt had not participated in the 1964 symposium because of some professional

The grand setting of the confluence of the Little Colorado and Colorado Rivers reveals that something of major proportions occurred here during the evolution of the river system. Photograph by Mike Buchheit

disagreements with a few attendees. In the thirteen years since his previous contribution, Hunt's thinking regarding the origin of the Grand Canyon had evolved, and his 1969 paper was published largely in response to the symposium's findings.

Hunt presented new ideas about how the placement and configuration of the Colorado River and its major tributaries may have changed through time. He wrote that the upper Colorado might have terminated in a lake basin in the Glen Canyon region about 25 million years ago. Because inflow did not exceed evaporation rates, there was no outlet from this lake. Hunt postulated that the ancient San Juan and Little Colorado Rivers once flowed in channels farther south than they do today and that they may have joined at some place west of the Kaibab upwarp. Sometime before 6 or 5 million years ago, the Glen Canyon lake basin overflowed its lowest rim and joined the two other streams to create something similar to the modern Colorado River.

As a possible explanation for the lack of recognizable Colorado River deposits in the Muddy Creek Formation, Hunt postulated that river water might have percolated into the subsurface near Peach Springs Wash and emerged from springs along the face of the Grand Wash Cliffs (background). These springs may have fed a lake that deposited the Hualapai Limestone (foreground). Photograph by Wayne Ranney

These interpretations were novel, some might even say forced, as few if any future geologists followed up on them to any degree. Arriving finally downstream at the Muddy Creek problem, as all Grand Canyon geologists eventually must, Hunt stated:

> *I postulate that the ancestral Colorado River (that is, the ancestral San Juan and Little Colorado) left the Colorado Plateau via the dry canyon at Peach Springs.*

This was an interesting idea resurrected in part from his 1956 paper. Geologists at this time were intrigued by this possibility, but since it was first proposed no one had actually found any evidence for the presence of Colorado River gravel in Peach Springs Wash. To fend off this criticism, Hunt noted that the river's pre-dam sediment load contained only 5 percent of rock types eroded from upstream of the Grand Canyon. To him, this could explain why no evidence was found for the river going south in Peach Springs Wash. But couldn't that also explain the lack of evidence of a Colorado River in Muddy Creek time? It seems Hunt was searching desperately to accommodate the ever-strengthening view that the river could not have been flowing into the Muddy Creek basin while it was being filled with sediment between 16 and 6 million years ago.

Instead, Hunt introduced a seemingly outrageous hypothesis to explain the apparent lack of any recognizable Colorado River gravels in the Muddy Creek Formation. He suggested that after leaving Peach Springs Wash to the south, the river became ponded, and water subsequently percolated into the ground. Limestone below was dissolved into the calcium-rich water, which was piped underground to the northwest, eventually issuing in springs from the walls of the Grand Wash Cliffs. This water, according to Hunt, fed the Hualapai Lake and deposited the Hualapai Limestone.

> *Water . . . could have discharged through the cavernous limestone to supply springs . . . [at] the mouth of the Grand Canyon where the maximum deposition of the Hualapai Limestone is centered. Such a mechanism involves piping on a truly grand scale. Some may believe the scale is outrageous, yet . . . the postulated piping would provide the large quantity of water required for the lake and the limestone deposited within it and would explain the absence of [Colorado River] deposits in the lake.*

Hunt was deeply troubled by the Muddy Creek problem since he saw evidence for the river being older upstream in the state of Colorado. He was also ambivalent about a polyphase history for the river proposed at the MNA symposium five years earlier. He hypothesized that it was just as easy to explain the river's confusing age by invoking the presence of a lake in the Glen Canyon area instead of a river flowing toward the Rio Grande since deposits for both scenarios are lacking. His ideas that the Colorado River experienced certain ponding episodes aligned him to some degree with Blackwelder's views, but the two men did not agree on the specific timing of these lakes, and Hunt's ideas were roundly criticized in the latter part of his career. No one has found evidence that the Colorado River exited through Peach Springs Wash. Surprisingly, Hunt's ideas regarding underground piping of water have been recently resurrected, and data has been collected for this hypothesis using more advanced techniques.

A Young Colorado River

© RAEchel Running.com

Ivo Lucchitta, 1972

Dr. Eddie McKee recruited two doctoral candidates to study the geology of the southwestern Colorado Plateau where it adjoins the Basin and Range. The two students, Ivo Lucchitta and Dick Young, presented the results of their work at the 1964 MNA symposium; these results may have been the driving impetus for Eddie McKee to call for the symposium since the new evidence spoke loudly to the origin of the Colorado River in Grand Canyon. Lucchitta studied the Muddy Creek Formation at the base of the Grand Wash Cliffs and went on to a stellar career revolving around the Grand Canyon and the southern Basin and Range. During the 1970s and 1980s, he became an articulate spokesperson for the evolution of the Colorado River in Grand Canyon.

His take on the Colorado's history was firmly planted in the "young river" camp, and he has been a vocal proponent of river integration through headward erosion and stream piracy. Lucchitta's conclusions mirror those of Longwell: he agreed that the Muddy Creek Formation contains no Colorado River sediment and that this supports the idea that the river and canyon must be younger than 6 million years. Lucchitta also revived the earlier proposal of Babenroth and Strahler, who theorized that the Colorado River established its course across the Kaibab upwarp in a subsequent valley between the diving nose of the upwarp on the north and a cliff of retreating Mesozoic formations on the south. Lucchitta called this the "racetrack theory," since the shape of the subsequent valley mimics a racetrack confined between the grandstand wall and the infield of a racing track (see illustration on page 74). He said this drainage flowed northwest away from the Grand Canyon area after leaving the area of the Kaibab upwarp to the west. He proposed that headward erosion by the Hualapai drainage captured the ancestral upper Colorado River near Kanab Creek west of the Kaibab upwarp. This was a variation on the findings from the 1964 symposium, which stated that stream capture occurred east of the upwarp, near the confluence of the Colorado and Little Colorado Rivers. Lucchitta also became involved with discussions regarding the origin of the Bouse Formation, which will be discussed later.

Northeast Drainage

Photograph courtesy of Richard Young

Richard Young, 1978

Dr. McKee also recruited Dick Young in the 1960s to study deposits found in some side canyons on the Hualapai Plateau. Young's work provided supporting evidence for an initial northeast drainage across the southern Colorado Plateau. He found gravel deposits lying within Hindu and Milkweed Canyons containing clasts that could only have been derived from the south of Grand Canyon. Textures within the gravels show a northeast transport direction, and Young suggested the gravels were stripped from the flanks of the Mogollon Highlands and deposited within the paleocanyons, which trend east and appear to connect with Peach Springs Wash. He named these gravels the Music Mountain Formation. A later period of gravel accumulation, called the Buck and Doe Conglomerate, was determined to be mainly of local origin only.

Some of Young's subsequent work on the Music Mountain Formation reveals the unroofing history of the Mogollon Highlands, whereby the topmost (youngest) bedrock layers were eroded from the mountains first and consequently laid down as the bottommost gravels. As the mountain wore away, older bedrock clasts buried the initial deposits. The evidence also suggested that the Rim gravels (the name comes from the Mogollon Rim near which the gravels were first described) are Paleogene in age rather than the younger age reported at the 1964 symposium. Young's work also differentiates how the two gravels formed in humid versus arid environments. The Music Mountain Formation contains a reddish matrix that is highly oxidized, while the Buck and Doe Conglomerate has a whitish matrix that is not oxidized. Young has correlated this difference with a worldwide climatic change during the Paleogene known as the thermal maximum event.

A Path around the Muddy Creek Problem

Courtesy of University of Texas at El Paso Library, Special Collections

Earl Lovejoy, 1980

By now it should be evident that there are essentially two key sticking points in establishing a coherent theory on the evolution of the Colorado River: the date of the "birth" of the Colorado River and the timing of the uplift of the Colorado Plateau. We cannot say when the river was born until we define what it is, but for now let's strictly define it as the river we see today—fully integrated and flowing westward in its present course across the Kaibab upwarp and then south toward the Gulf of California. Using this definition of the river, the presence of the Muddy Creek Formation, which does not contain recognizable Colorado River sediment and formed between 16 and 6 million years ago, means that the Colorado River could not have been born earlier than 6 million years ago. Or could it?

Earl Lovejoy devised a way that the river could be older than 6 million years and still exit the Grand Wash Cliffs during Muddy Creek time. Ironically, he used the very same Muddy Creek deposits that Longwell and Lucchitta had used before him to preclude the presence of the river here. His model was an analogous example from the Rio Grande in Big Bend National Park, Texas. There, after exiting Santa Elena Canyon, the river is forced against the base of a cliff by the progressive growth of a large alluvial fan that approaches the river from a nearby mountain. Lovejoy proposed a similar setting for the Colorado River during Muddy Creek time, postulating that the river's course may have been forced to turn abruptly to the south after emerging from the Grand Wash Cliffs. He reasoned that such a sharp deflection was accomplished by the progressive accumulation of Muddy Creek sediment toward the Grand Wash Cliffs from west to east. He explained that this might have confined the river to a course that was narrowly set against the base of the cliffs and could reasonably explain the lack of definitive Colorado River sediment within the bulk of the Muddy Creek Formation.

According to Lovejoy, if this scenario were true, recognizable Colorado River sediments should be found exactly where the river is suspected to have flowed—along the base of the southern Grand Wash Cliffs. There are significant sand and silt deposits within the Muddy Creek Formation here but no gravels. Lovejoy theorized that these sands and silts were indistinguishable from modern Colorado River sediment. Concerning the admitted absence of more diagnostic Colorado River gravel, he used an example from the Truckee River near Reno, Nevada, where diagnostic river gravel was present in an upstream reach of that river but had settled and therefore was removed from the river's load above a narrow canyon such that only sand and silt were deposited downstream.

Lovejoy further suggested that the Colorado could have become ponded beneath the Grand Wash Cliffs, forming the Hualapai Limestone that today rests on top of coarser gravels of the Muddy Creek Formation. He noted that as this basin progressively filled with water, the lake probably extended far up into an enlarging Grand Canyon. According to him, all of the coarser debris in the river was deposited at the head of this lake far upstream in Grand Canyon. Because of this, only clear water exited the Grand Wash Cliffs, thus explaining the origin of the Hualapai Limestone. Eventually, as this basin became filled with algae-derived limestone, spillover of the lake established the modern course of the river to the west toward the Hoover Dam area. Ignoring the more dominant views of the time, Lovejoy accordingly established a viable alternative solution to the Muddy Creek problem.

Lovejoy generally favored the ideas of Powell, Dutton, and Hunt, who envisioned an old Colorado River that flowed west; he was less influenced by the more modern views of Blackwelder, Longwell, and Lucchitta,

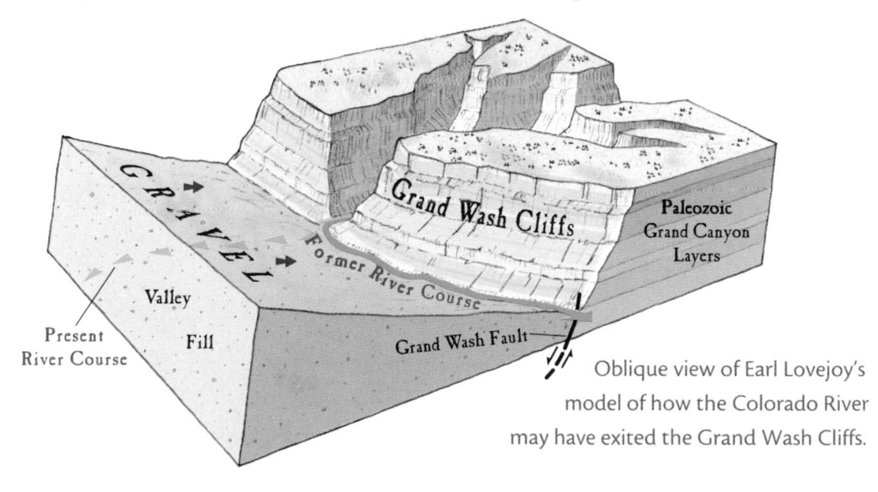

Oblique view of Earl Lovejoy's model of how the Colorado River may have exited the Grand Wash Cliffs.

who thought the river was young. In order for his theory to be valid, he appealed for an older age for the Basin and Range than is usually ascribed, since the river could only flow to the west if there was a topographic low in that direction. The accumulating evidence, however, suggests that highlands, not lowlands, were present to the west of Grand Canyon until at least 17 million years ago, and these highlands directed river flow toward the northeast. So it's unlikely, as Lovejoy postulated, that an old Colorado River went west in early Cenozoic time.

Lovejoy's observations from modern river settings in Texas and Nevada offer an alternative way to answer the Muddy Creek problem and how the Colorado River may have exited the Grand Wash Cliffs without leaving behind telltale gravels. His is an intriguing idea that is mostly ignored by modern geologists. Nonetheless, it reveals to us the ways in which geologists can find alternative explanations to a specific problem when there is seemingly irrefutable evidence to the contrary.

An Old Grand Canyon

Courtesy of Karen Elston Davis

Don Elston, 1991

As our discussion nears the end of the twentieth century, a final stab at an argument for an old Grand Canyon roared back to life. Don Elston, a geologist with the U.S. Geological Survey, arrived on the Colorado Plateau in the 1950s to map uranium deposits, which were used in the Cold War effort. However, his personal and professional association with Charlie Hunt soon had him thinking about the origin of the Colorado River.

Elston saw evidence in the landscape for a period of deep dissection that may have commenced as early as 100 million years ago when, according to him, the plateau was tilted toward the north. He thought that the Grand Canyon was cut to its present depth sometime between 70 and 60 million years ago by a river that flowed to the northeast. Elston postulated that the canyon was completely buried later by the Rim gravel, which spread north from the Mogollon Highlands. Subsequently, as the Rim gravel was eroded, the Grand Canyon was exhumed. Then the modern Colorado River was established as the plateau reversed its tilt (as

Davis proposed in 1900), this time down to the south and west, causing the river to occupy the same channel but flow the other way. Important to this idea is the notion that the Grand Canyon was carved to its present depth between 70 and 60 million years ago.

Elston further stated that by about 10 million years ago hyperarid climatic conditions caused the Colorado River to become seasonal, and its channel became choked with local slope deposits to a depth of about one thousand feet. The damming of the river by these gravels explained the lack of Colorado River deposits in the Muddy Creek Formation. According to this idea, the river flowed in the subsurface beneath the gravels through Grand Canyon toward the Grand Wash Cliffs. Wetter conditions later allowed the Colorado to re-excavate its channel.

The idea of a Paleogene origin for the Grand Canyon is not embraced by the majority of geologists, who insist that the age of the river (at 6 million years) necessarily implies a specific age for the Grand Canyon. Although the idea of an old canyon does not represent a majority view, theories such as Elston's help us to define the outer limits for when the river could have evolved to its present configuration.

We are beginning to see that deciphering the history of the Grand Canyon is similar to the story of the six blind men who describe the elephant they can touch but cannot see. Each man describes an animal based on the part he happens to touch, but taken together, they seem to be describing six different animals.

So the twentieth century closed without a widely accepted theory on Grand Canyon's origin. Older ideas were occasionally rehashed and reworked, but newer interpretations were also presented. Much of this was about to change, however, with another conference that would bring geologists together to consider the possible origin of the Colorado River and the Grand Canyon.

THE TWENTY-FIRST CENTURY

Studies related to the origin of the Grand Canyon and the Colorado River languished somewhat in the final one-third of the twentieth century. By the time the new millennium began, thirty-six years had passed since the symposium at the Museum of Northern Arizona, and it remained the only scientific gathering convened specifically to discuss the canyon's puzzling origin. To frame the intellectual setting of the day, we must recall that this meeting took place before the fundamental theory of plate tectonics was widely accepted (generally around 1968), a truly astonishing fact in hindsight. In addition, radiometric dating techniques were in their infancy. Regardless, a significant idea did emerge from the symposium in 1964, namely that the Colorado River likely formed from the integration of two or more distinct drainage systems. Although some of the specific details regarding that theory were later proved untenable, the larger contribution was that it showed how rivers could evolve from multiple ancestors.

The 2000 Symposium

The dawn of the new millennium saw a renewed interest in research related to the origin of the Grand Canyon. In June 2000, a professional symposium was held at Grand Canyon National Park with seventy-three geologists in attendance and thirty-six scientific abstracts submitted for review. Attendees on a pre-meeting field trip went to the Bidahochi Formation north of Holbrook, Arizona, and the Rim gravel on the Mogollon Rim. Geologists on a post-meeting field trip looked at gravel deposits in Milkweed Canyon on the Hualapai Plateau, and in the Grand Wash trough. A scientific monograph titled "Colorado River: Origin and Evolution" was published by the Grand Canyon Association in 2004 and included articles written for the most part by symposium participants, for use by professional geologists.

Summarizing the broad range of ideas presented at the 2000 conference is inherently subjective since every participant observed the conference through the filter of their own experience. New ideas are sometimes met with vigorous resistance because geologists tend to be skeptics first and grow to be supporters only when the evidence

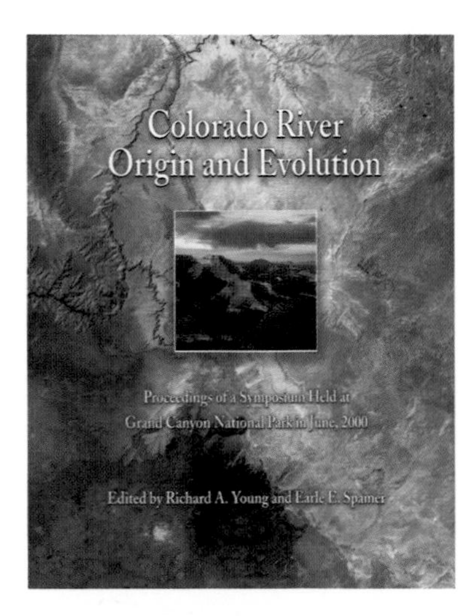

"Colorado River Origin and Evolution," 2004. Courtesy of Grand Canyon Association

is strong. No matter how one interprets the conference's results, new ideas did enter the debate, and they reinvigorated interest regarding the origin of the river and the canyon.

Four broad categories summarize the theories discussed at the 2000 meeting:

1. The role of headward erosion and stream piracy in shaping the river
2. An alternative theory of basin spillover associated with ancient lakes
3. Possible reverse flow of the river in all or parts of Grand Canyon
4. Significant recent deepening of the canyon

During the symposium, the atmosphere was charged between the supporters of headward erosion and those for spillover. Geologists in the headward erosion camp expected a strong challenge from spillover advocates, and neither group was denied. They each agreed that the Colorado River was likely integrated from two or more separate and distinct drainage systems—their differences lay only in the specific process that did the integrating. Reverse flow for the Colorado River's entire length in Grand Canyon or only for various portions of it was debated. Proponents of the entire river being reversed in Grand Canyon did not need to invoke an integration event, while those supporting reversal of only selected parts did. Advocates for recent deepening in only the last few million years proposed that this deepening likely resulted from climate change (and the subsequent increase in river discharge) or was facilitated by the differential uplift along faults in the western canyon, or both.

Headward Erosion and Stream Piracy

In the lead-up to the 2000 symposium, some geologists began to question the efficiency of headward erosion in acting upon arid-land river systems. They pointed out that the amount of runoff in the rather limited headwall area of a steep-gradient stream might be too minimal or deficient to cause it to lengthen its channel. At the beginning of the symposium, a buzz was in the air that headward erosion would be seriously challenged, but the old guard came prepared to defend it and was armed with fresh insights and new evidence.

Ivo Lucchitta was one of the leading defenders of a Grand Canyon formed by headward erosion in the last 6 million years. He recalled having mapped the top of the Shivwits Plateau in western Grand Canyon, where gravels containing certain clast types convinced him the deposits could only have come from south of the canyon. Since the gravels were now exposed on the high-standing Shivwits surface and were capped by a 6-million-year-old lava flow, he indicated that the Grand Canyon could not possibly have existed prior to this time—how else could the gravels travel from south to north across a canyon? Lucchitta was also in favor of a stream capture point between the Shivwits and Kaibab Plateaus near Kanab Creek. This was a significant modification of the capture point suggested by McKee at the 1964 symposium.

Jim Faulds reported on mapping he completed below the Grand Wash Cliffs at the boundary between the Colorado Plateau and the Basin and Range. He obtained age constraints (a range of ages) on the movement of the Grand Wash Fault and showed that this faulted escarpment appeared on the landscape between about 16 and 13 million years ago. At this time, he submitted, headward erosion possibly began carving into the western edge of the Grand Wash Cliffs. The finding seemed to refute one of the major arguments against headward erosion; namely, the abrupt appearance of Colorado River gravel in this area at about 6 million years ago necessarily meant that headward erosion began rather instantaneously at this time. Faulds proposed that headward erosion back into the plateau edge might have operated over a much longer period that culminated with an integration event about 6 million years ago.

A graduate student working with Dick Young also showed how West Clear Creek on the Mogollon Rim in central Arizona most likely eroded headward in capturing portions of East Clear Creek on the other side of the drainage divide. The study was undertaken to show that evidence exists for headward erosion acting upon the edge of the Colorado Plateau, even on a stream much smaller than the Colorado River. According to the results of this study, headward erosion has occurred here in the last 10 million years and has progressed back into the plateau edge by more than twenty miles.

Although some geologists came to the 2000 symposium with serious doubts about the efficiency of headward erosion forming the Grand Canyon, it became obvious that the idea was not going to die at this time. Supporters pointed out that any suspected inefficiencies with the process might simply be from the mistaken assumption that the process

The Grand Wash Cliffs, which mark the boundary between the Colorado Plateau and Basin and Range Provinces, were formed by faulting beginning about 13 million years ago. Photograph by Wayne Ranney

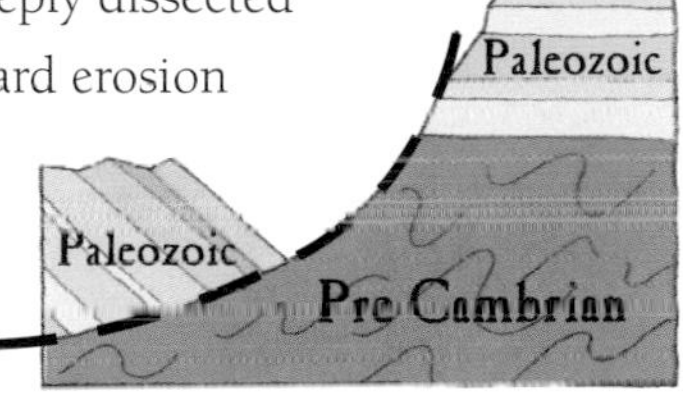

of headward erosion led simultaneously to the deeply dissected canyon we see today. This is not the case. Headward erosion and stream capture served only to integrate two distinct drainages and presumably upon a very subdued terrain. Different processes then worked to deepen the canyon after integration, making it the profound chasm that graces the landscape today.

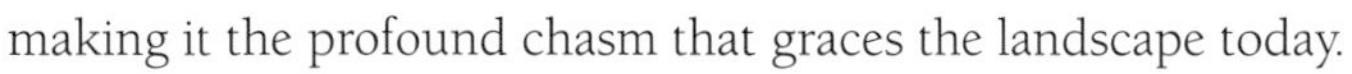

Spillover Theory

The process that challenged the status of headward erosion at the 2000 conference was closed-basin spillover, whereby interior drained basins "filled and spilled" to integrate the Colorado River. Some geologists were convinced that a large Lake Bidahochi existed east of the Grand Canyon about 6 million years ago and that it may have drained rapidly (geologically speaking) in establishing the modern river. Others remained skeptical, and the debate was of such importance that it demanded a field trip to the Bidahochi Formation to view the evidence. Other attendees focused on deposits downstream of the Grand Canyon that

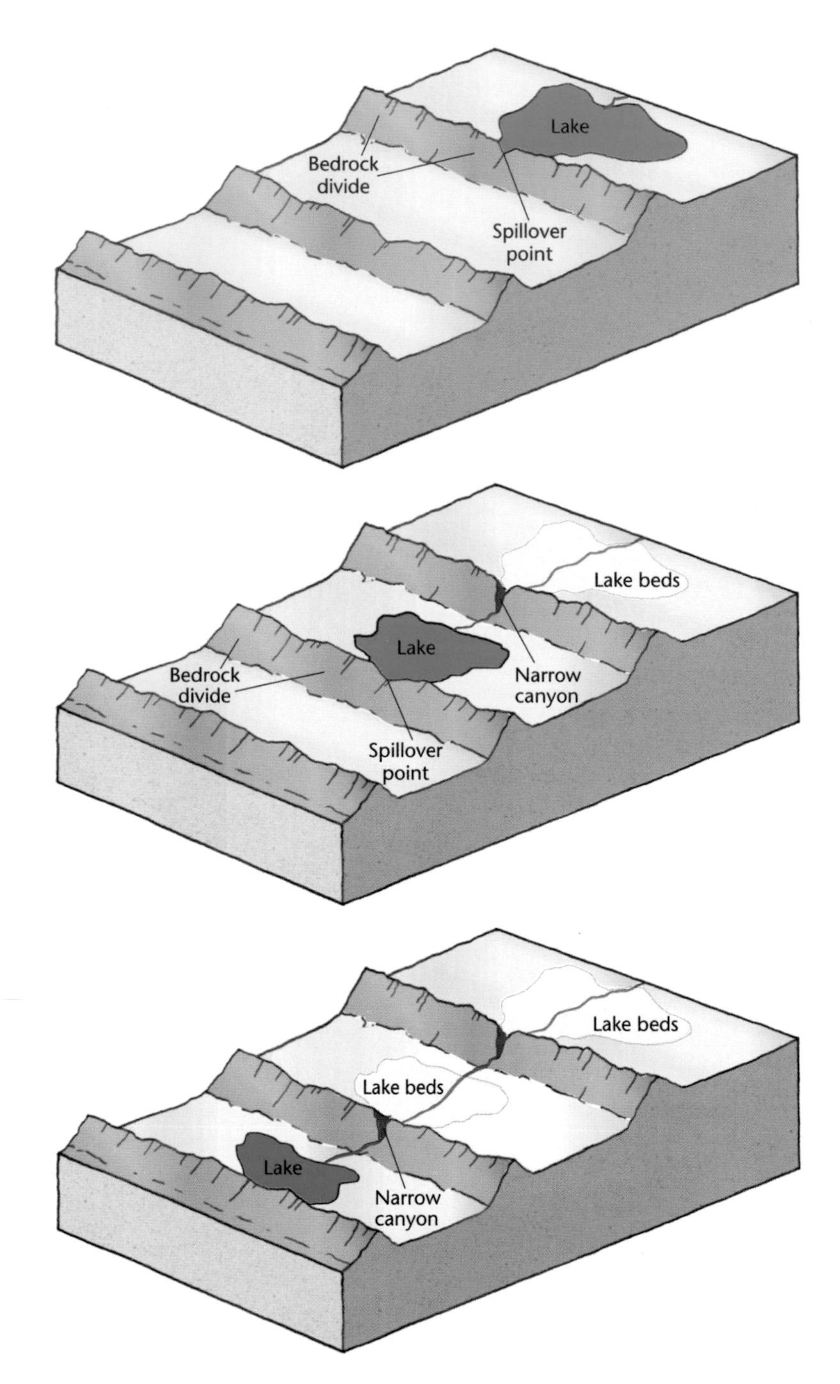

Simplified diagram showing how sequential basins fill with water and spill over bedrock divides to form a river that alternately flows in open valleys (former basins) and narrow canyons (former divides).

spoke to evidence for spillover on this part of the river. Much of this effort was an attempt to clarify the controversial origin of the Bouse Formation as either marine or freshwater. At the 2000 symposium, spillover theory was nudged to the forefront of ideas regarding the origin of the Colorado River.

Recall that Blackwelder was the first geologist to suggest that the river formed by this process but that he did not offer evidence from river deposits. Rather, he formulated his idea by recognizing the curious fact that the river today flows alternately through wide, open basins (where the lakes would have been) and in narrow canyons (where the lakes overtopped and spilled). This was a fantastic observation that hinted at the river's young age. Geologists at the 1964 symposium were favorably inclined toward a large Lake Bidahochi, but at the 2000 conference, the pre-meeting field trip resulted in mixed reviews about whether the deposits represented an expansive lake or were merely the scattered remnants of disconnected or ephemeral ponds.

Norman Meek and John Douglass came to the defense of a large Lake Bidahochi and envisioned that it spilled rather rapidly, following an already established line of drainage through the future Grand Canyon. They used modern analogies from elsewhere in the Southwest such as how the Mojave River in California rapidly overflowed a basin to establish its course. They proposed that a basin in central Utah overflowed and filled Lake Bidahochi by about 6.5 million years ago. Then Lake Bidahochi spilled about 5.5 million years ago, perhaps at a low spot that once existed between Desert View on the South Rim and Cape Royal on the North Rim, thus initiating the incision of the Grand Canyon. (Curiously, Meek came to his interpretation without being previously aware of Blackwelder's hypothesis.) According to Meek and Douglass, spillover theory could explain how the Grand Canyon was carved without invoking a period of recent plateau uplift. Some geologists were fond of saying, "No uplift, no canyons," but Meek and Douglas modified it to say, "No base level change, no canyons." They offered a fairly nuanced view of canyon cutting that showed that a change in base level might only make uplift apparent.

Meek and Douglass clarified to skeptics that the draining of Lake Bidahochi need not create its own line of drainage but likely followed a preexisting channel. It was interesting to speculate where their

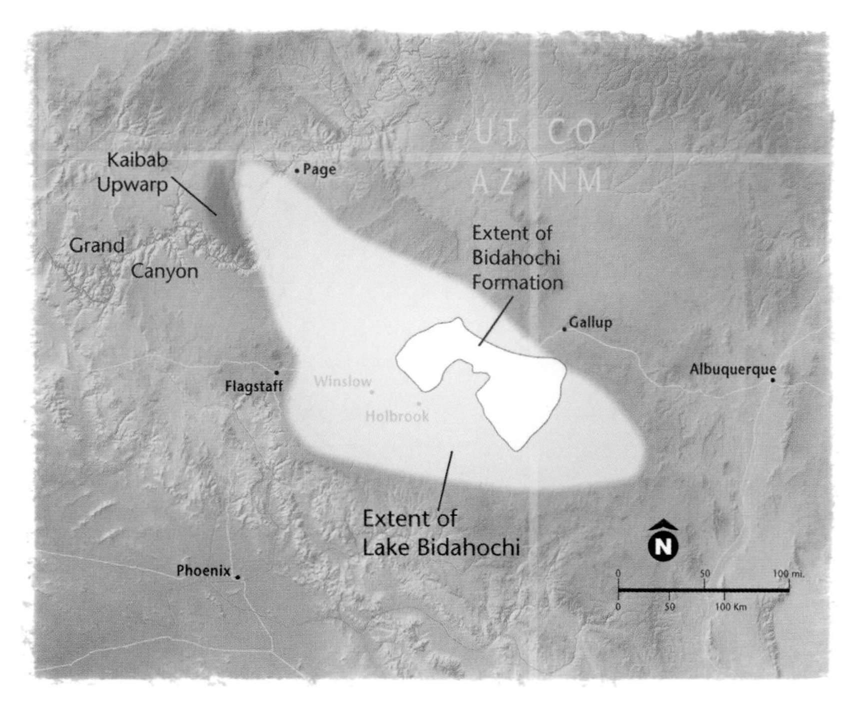

Regional map showing the possible former extent of Lake Bidahochi in northeastern Arizona. The present-day elevations of the Bidahochi Formation remnants allow geologists to postulate the lake's possible size.

proposed spillover would have gone—could it have been toward or into the drainage that Jim Faulds suggested was eroding headward back into the plateau edge? Remember he proposed that headward erosion began no later than 13 million years ago. It is tantalizing to combine the two ideas since it is plausible and (ironically) incorporates two seemingly contradictory processes as coeval mechanisms to create the modern river.

Jon Spencer and Philip Pearthree submitted a tough critique on the inefficiency of rapid headward erosion acting in the arid Southwest, proposing that closed-basin spillover was a more likely process to have integrated the river. They proposed (as Blackwelder had previously) that the Colorado River might have flowed from its lofty headwaters into lakes located on the central Colorado Plateau. They noted that the rate of sediment accumulation at this time (about 15 million years ago) was only one-third of present-day rates, and this could possibly explain the apparent lack of evidence for large lakes. They speculated that Lake Bidahochi could have filled rather rapidly, then spilled to the west to integrate the river through Grand Canyon, concluding that a short-lived Lake Bidahochi would not leave much evidence for its brief existence. They offered this as an alternative to headward erosion, which to them was unfathomable in this setting.

Jonathan Patchett and Jon Spencer provided preliminary evidence for a spillover origin on the lower Colorado River in the Mojave Desert. Here the river flows through multiple basins separated by narrow canyons cut into solid bedrock. They reported that within the broad basins are isolated outcrops of the Bouse Formation that are found at various elevations, with some as high as eighteen hundred feet above sea level. Some geologists had previously explained these differences in elevation as the result of recent faulting and uplift of the Colorado Plateau breaking across formerly continuous marine deposits of the ancestral Gulf of California. But Patchett and Spencer noted that no faults could be found between the various outcrops and suggested instead that the Bouse was deposited in freshwater lakes that were integrated through closed-basin spillover.

The specific environment of deposition of the Bouse Formation was controversial at the 2000 symposium, and a clearer understanding of its origin held great importance to both sides of the debate. If the Bouse deposits were marine in origin, then the present-day outcrop pattern would be evidence for a period of recent plateau uplift (the deposits are around 5 million years old). If the deposits formed in multiple freshwater lakes, then recent plateau uplift would be unnecessary to explain the discrepancy in outcrop elevation. To add to the confusion, the fossil evidence was contradictory. Patchett and Spencer offered chemical data that showed evidence for a freshwater origin, thus supporting closed-basin spillover for this part of the river. We will hear more about these deposits when we examine some results from the 2010 workshop.

Reversed Flow of Colorado River

The notion that the Colorado River reversed its flow direction was one of the major topics discussed at the 2000 symposium. Previous evidence supported the concept of a regional northeast drainage system in the

The Bouse Formation, shown here in an outcrop near Cibola, Arizona, is an important indicator of how the lower Colorado River may have evolved from basin spillover. Photograph by Jon Spencer

early Cenozoic that was opposite to the present-day flow direction of the Colorado River. More-specific support was submitted at this conference showing how a precursor to the Colorado River in Grand Canyon might be related to the northeast drainage. This necessitates a later period of drainage reversal for the river, and the timing and possible process that accomplished the reversal was debated. Additionally, the extent to which the modern landscape preserves elements of the ancient northeast drainage system enticed some to speculate upon that theory.

Dick Young reported on gravels that originated in a northeast flowing system across the Grand Canyon region. He mapped and described nearly twelve hundred feet of deposits located in Milkweed, Hindu, and Peach Springs Canyons, all on the Hualapai Plateau (see map on pages 18–19). The various formations located here provide evidence for northeast-directed flow, the gradual unroofing of the Mogollon Highlands, and a climate shift to more arid conditions by mid-Cenozoic time (about 30 to 25 million years ago). The oldest gravel, the Music Mountain Formation, was deposited in paleocanyons that are about four thousand feet deep and documents a significant period of uplift here in the early Cenozoic. A younger gravel deposit, the Buck and Doe Conglomerate, appeared to have completely buried the paleocanyons, and Young speculated that superposition on this surface may have positioned the course of the Colorado River here.

Andre Potochnik described similarities between the upper Salt River region in eastern Arizona and the Grand Canyon. He found that a Salt River paleovalley was cut to a depth of about forty-six hundred feet in the early Cenozoic by a stream that flowed northeast as recently as 33 to 30 million years ago. Drainage reversal occurred after 15 million years ago. Potochnik argued that the course of the Colorado River in Grand Canyon today might be an inherited relic from the period of northeast flow, analogous to what he documented in the Salt River paleovalley. Don Elston, you will recall, accepted this broad interpretation with the added condition that the present depth of the Grand Canyon was attained when the river flowed northeast. The 15-million-year date for drainage reversal in the Salt River paleovalley agrees in general with Faulds's finding that headward erosion had begun to carve its way eastward into the edge of the Grand Wash Cliffs.

An important question remained after the symposium in 2000: Does the Colorado River in Grand Canyon preserve or utilize any portion of the old northeast drainage system after it became integrated 6 million years ago?

Recent Deepening of the Grand Canyon

While the processes that may have integrated the river and formed the canyon were debated at the 2000 symposium, the rate at which the Colorado River might be deepening the Grand Canyon also received attention. The role that glacial meltwater may have played in deepening the canyon during the ice age was long ignored by scientists, and historically, very little attention was focused on the role climate played in carving the canyon. Only one climatologist attended the 2000 symposium, and he directed his interest only to a time about 6 to 4 million years ago and in the area of the Colorado River delta. The suspected increase in runoff for the river during the ice age would have created a greater erosive capacity for the river to carve the canyon (see chapter 3). This is especially true if climatic factors were given a boost by tectonic ones, and such a scenario was also presented at the conference.

Ice did not carve the Grand Canyon nor did it affect its shape or profile in any direct way—glaciers did not extend this far to the south. However, the Colorado River extends its reach far north into landscapes that were glaciated, meaning that glacial meltwater traveled through the Grand Canyon on many occasions during the last 2 million years. The episodic growth and melting of glaciers in the Rocky Mountains meant they advanced and retreated numerous times, alternately storing and releasing large volumes of water through the Grand Canyon.

Based on results presented by Cassie Fenton, Bob Webb, and Philip Pearthree, increased cutting capacity of the Colorado River likely was enhanced by recurrent movement on the Toroweap and Hurricane Faults in western Grand Canyon. These faults broke through volcanic rocks and alluvial fans that are between thirty thousand and four hundred thousand years old. The geologists then measured the total displacement of the two faults at about nineteen hundred feet and extrapolated that the bulk of this displacement could have occurred in the last 3.5 million years. Their idea suggests that as the western Grand Canyon was lowered, cutting of the eastern canyon accelerated.

Joel Pederson and Karl Karlstrom looked at the rates of displacement on the same two faults but concluded that these rates were insufficient to carve the present depth of the entire Grand Canyon. They further stated that the displacement rate on the Toroweap Fault is not large enough to drive upstream deepening of the canyon and that an earlier integration event at 6 million years may be a more important factor. Their conclusion states that the rate of canyon cutting has either diminished through the last 6 million years (as the river progressively lowered its gradient) or, alternatively, some of the canyon's present depth was achieved before integration of the river took place. They suggested that the incision of the Grand Canyon resulted from a combination of early Cenozoic uplift, the differential displacement between the Colorado Plateau and the Basin and Range, and the integration of the river off of the plateau edge.

Thomas Hanks, Ivo Lucchitta, and others looked at pediment surfaces found on the northern slopes of Navajo Mountain and reported when the gravel deposits on those surfaces were dissected. They used surface exposure ages garnered from clasts in the gravel deposits (see sidebar on pages 124–125). Their conclusion stated that Glen Canyon, now approximately

Rocky Mountain glaciers (such as Andrews Glacier), were a huge source of runoff during the ice age in Grand Canyon. Photograph by John Crossley, The American Southwest, www.americansouthwest.net

eight hundred feet deep (disregarding inundation by Lake Powell), was cut to its present depth in only the last five hundred thousand years.

All of these studies show that the relatively recent (4 million years or less) deepening of the Grand Canyon region was significant, about two thousand feet or more. The symposium in 2000 ended with geologists eager to return to the field and explore the questions raised during this important event.

The 2010 Workshop

In May 2010 a third meeting was convened at the U.S. Geological Survey in Flagstaff, with fifty-nine geologists attending. The workshop lasted three days and was aided by rapid advances in the use of electronic media since the 2000 symposium. Fifty abstracts were submitted to a website where geologists submitted and reviewed the findings of their colleagues ahead of the gathering. See the Scientific Bibliography (pages 181–184) for websites relating to the workshop.

While summarizing the many new ideas presented at this workshop may appear overwhelming, a look at the abstract titles will serve to show what ideas are becoming better known and what questions still puzzle geologists. No matter how one interprets the results from the 2010 workshop, the new ideas presented serve to invigorate the debate. In synthesizing the current state of knowledge, it is helpful to define six broad lines of research currently under way:

1. Uplift mechanisms of the Colorado Plateau
2. Evidence for early ancestors of the Colorado River
3. An old Grand Canyon
4. Lack of evidence for mid-Cenozoic drainage across the region
5. Karst connection and groundwater processes
6. Evidence for a young Colorado River

An examination of each of these can help us gain an appreciation for the most up-to-date theories on Grand Canyon's origin.

Uplift Mechanisms of the Colorado Plateau

Of the many enigmas related to the origin of the Grand Canyon, the timing and cause of Colorado Plateau uplift is its Achilles' heel. Much speculation revolves around when and how the plateau was uplifted simply because it is much harder to recognize evidence for an uplift event than it is for a deposit lying on the ground. New and leading-edge techniques, however, are helping geologists garner new information about the possible uplift history of the plateau.

Support for recent plateau uplift occurring in the last 6 million years comes from Karl Karlstrom, Ryan Crow, and others who determined that hot, upwelling mantle might be driving the surface uplift and volcanism at the southern edge of the Colorado Plateau in western Grand Canyon. The study uses computed tomography, a sort of CT scan that provides an image of the variable compositions and temperatures within the earth's interior. The procedure records the speed of seismic waves, which travel faster through the cooler outer lithosphere (composed of the crust and the rigid uppermost part of the mantle) and slower through the asthenosphere (the hotter, more pliable lower part of the mantle). Their research detected the shallow presence of a hot asthenosphere beneath western Grand Canyon,

Lava flows such as these above Whitmore Wash in western Grand Canyon may reveal much about the uplift history of the Grand Canyon region. Photograph by Michael Collier

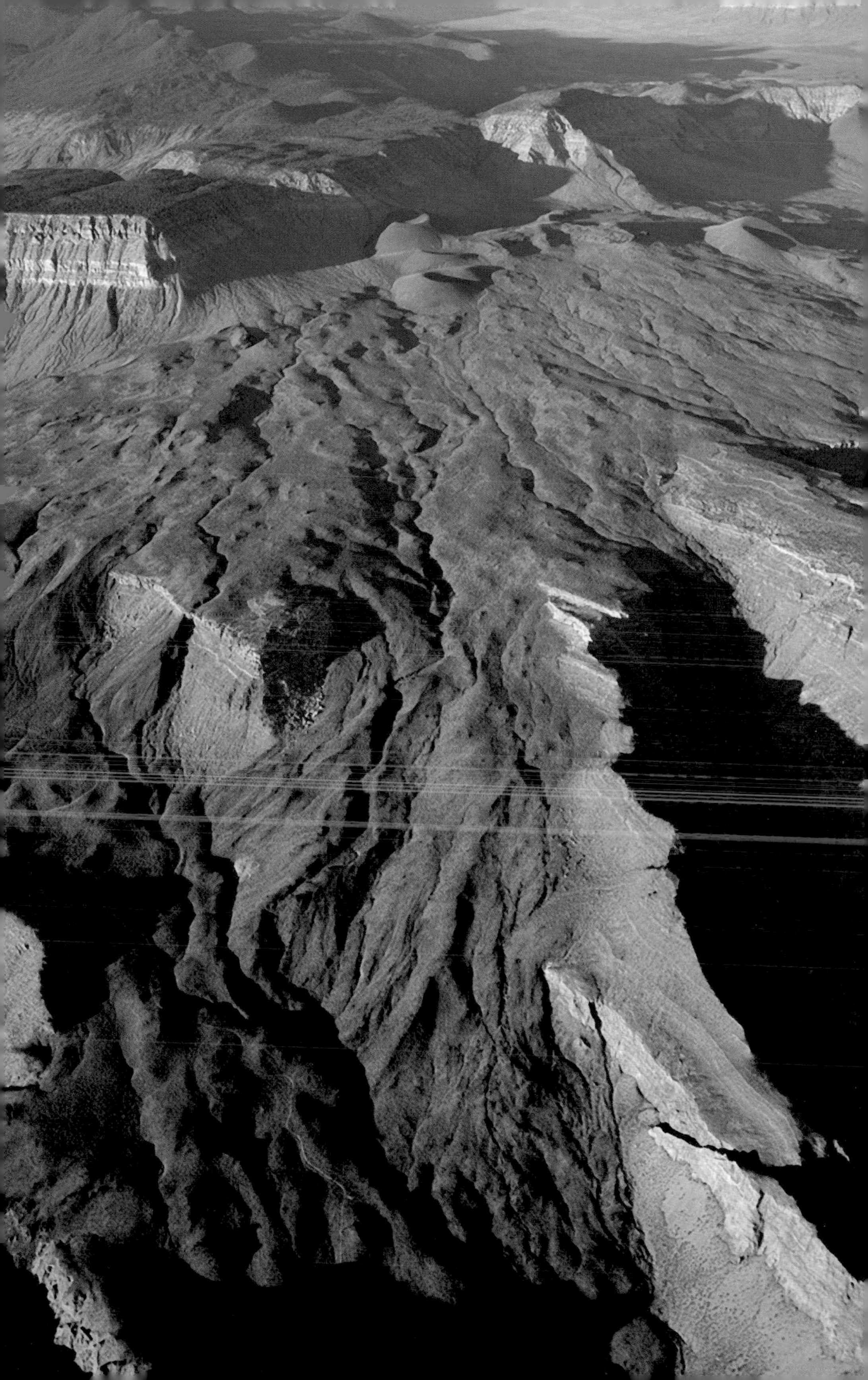

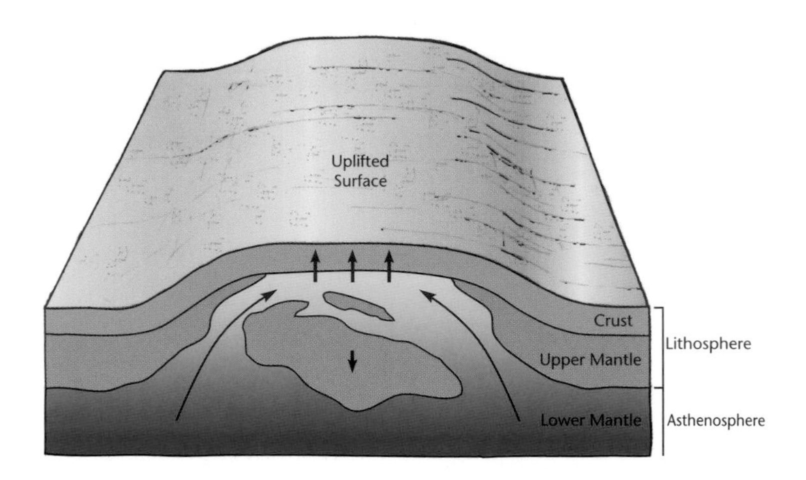

The mantle portion of the earth's lithosphere (shown in blue) may have become detached from the crust (shown in green), which allowed the hot lower mantle (orange and yellow) to rise up and replace it. This could have driven recent uplift of western Grand Canyon.

and a correlation is inferred with areas suspected of experiencing active uplift. They explained that the mantle portion of the lithosphere might have become detached from the crust and gradually dripped into the lower mantle. When this happened, the asthenosphere invaded the area beneath the crust, heating it and driving the uplift of the plateau's edge.

Recent movement on the Toroweap and Hurricane Faults might be the surface expression of this buoyant rise of the asthenosphere beneath western Grand Canyon. The invasion of hot, buoyant asthenosphere would cause the crust to become elevated and stressed, activating fault movement. The authors proposed a possible connection between this and the observed high gradient of the Colorado River in Grand Canyon (relative to other portions of the river)—an actively uplifting terrain would cause an obstacle to form across the river's path and force the river to slice more forcefully through it. The use of this computed tomography provides additional support for the role that recent uplift may have played in creating the Grand Canyon's depth.

Regarding new evidence for an earlier period of uplift, Shari Kelly, Karl Karlstrom, and others used apatite fission track dating (see sidebar on pages 124–125) to provide evidence for multi-stage erosion intervals during the Laramide, mid-Cenozoic, and late Cenozoic time periods. Their study showed how deeply buried rocks on the Colorado Plateau and Rocky Mountains were cooled as erosion removed the overlying (and insulating) material during uplift. They reported that the area of the Mogollon Highlands in southwestern Arizona had two and a half miles of sedimentary rock removed from it during Laramide uplift. In eastern Grand Canyon, they found that up to one mile of additional rock strata covered the Kaibab Limestone until mid-Cenozoic time. Finally, they stated that the central Colorado Plateau was exhumed only 7 to 6 million years ago, although they did not determine whether this rapid erosion was exclusively from uplift of the area or from the integration of the river with attendant dissection of the landscape. Thermal-cooling techniques have been a windfall in resolving the uplift history of the region, and we can expect more surprising results from this area in the future.

Another type of uplift, called isostatic uplift, involves a passive response when erosion removes higher rock layers. On average, the Colorado Plateau has been uplifted about seven thousand feet since the sea last left the area about 80 million years ago. As relatively young marine rocks were stripped off the landscape, the confining pressure they exerted on the underlying rocks was removed, causing deeper rocks to rise, or float, upward. This isostatic uplift is passive since it is not a direct result of tectonic mountain building, although the initial cause of the denudation may or may not be related to tectonic uplift. Joel Pederson has shown that there are more potential sources of uplift on the Colorado Plateau (Laramide compression, mantle-driven uplift, and isostatic uplift) than there is actual uplift, meaning that some source (or sources) of uplift may be overstated in their importance. Further research will undoubtedly help to rectify the relative importance of each mode and period of uplift.

Evidence for Early Ancestors of the Colorado River

Geologists accept the assertion that the Colorado River can be no older than about 80 million years because prior to this time the region was beneath sea level. This date is not controversial among geologists and provides the maximum age for the river and thus the canyon. When the sea began to move away from the Grand Canyon region, a nascent drainage pattern developed that flowed toward the retreating shoreline to

The Laramide Orogeny and Other Periods of Plateau Uplift

Soon after John Wesley Powell proposed antecedence for the origin of the Green and Colorado Rivers, evidence began to accumulate that the uplifts on the Colorado Plateau might be older than the rivers. Geologists named this early uplift event the Laramide Orogeny (after deposits found in the Laramie Basin in Wyoming), and its importance in the Grand Canyon story is paramount. Evidence in central Arizona reveals a second period of uplift, occurring between 35 and 25 million years ago, called the Mid-Cenozoic Orogeny. And as some geologists sought to explain the apparent youthfulness of the canyons on the Colorado Plateau, they invoked a late Cenozoic, or recent, period of uplift. Consequently, three periods of uplift have been proposed to explain the uplift history of the plateau landscape—the Laramide or early Cenozoic, the mid-Cenozoic, and a recent, or late Cenozoic, event.

As individual geologists scrutinize the various lines of evidence for each of the uplift events, they naturally tend to favor the importance of some events over others. For this reason, the relative importance given to any

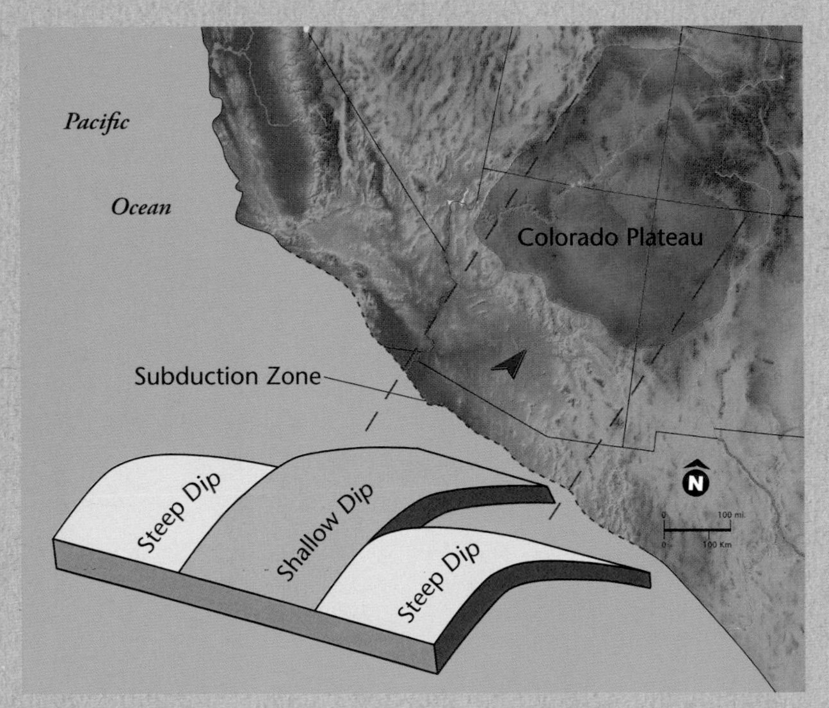

The Farallon Plate was likely subducted beneath western North America in numerous segments, which slipped below the continent at variable angles. The unique landscape of the Colorado Plateau relative to its neighbors may be the result of a slab dipping at a shallow angle.

Portions of western South America may be a perfect modern analog for the tectonic and geomorphic setting of the Colorado Plateau and Grand Canyon region of approximately 60 million years ago. Courtesy of NASA Goddard Space Flight Center, http://visibleearth.nasa.gov/

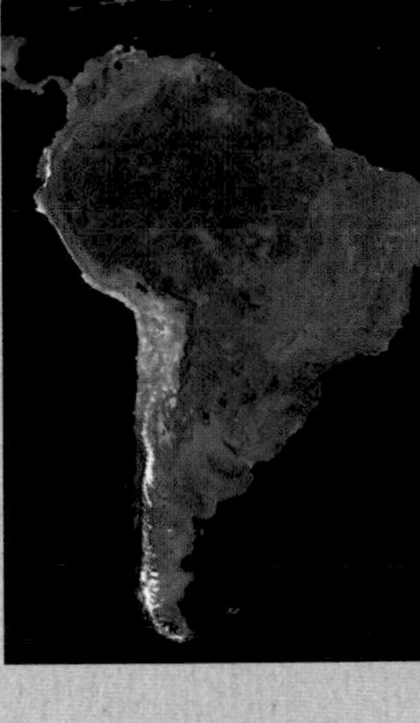

uplift episode varies: Laramide uplift seems to hold sway with some geologists, evidence for the Mid-Cenozoic Orogeny is favored by others, and the late Cenozoic has its many adherents as well. While all three uplift events likely played some role in helping to shape the modern plateau landscape, the mechanisms that drove each uplift event are what grab the attention of some geologists.

The Laramide Orogeny occurred between 70 and 40 million years ago and uplifted the Rocky Mountains in Colorado and Wyoming, further elevated the existing Mogollon Highlands in southwestern Arizona, and raised the Colorado Plateau to some unknown elevation. What is unique about the Laramide is how far inland the deformation migrated relative to the position of the plate boundary in far-off California. Geologists seem to be gaining ground in understanding why this happened.

In explaining the puzzling distribution of uplift, geologists note that oceanic plates are not simple, broad sheets of crust that are subducted uniformly along plate boundaries for thousands of miles. Rather, ocean plates tend to be subdivided into smaller segments with widths typically around five hundred miles or so. Some segments move faster or slower relative to their neighbors, but generally movement occurs at the speed that fingernails grow—about one to two inches per year. Each segment is bounded by lateral faults that accommodate the variable speed of each slab. Important to the discussion here is that these segments may also subduct at varying angles beneath the continent's edge, with some segments dipping rather steeply while others dip at a more shallow angle. Deeply dipping oceanic segments become heated much closer to the continent's edge, producing a line of volcanic mountains parallel to the shore—this is the setting today in much of western South America and the Andes Mountains. Curiously, not all parts of the Andes have active volcanoes, and geologists suspect that these parts of the range correlate closely with segments of the ocean plate that are thought to be dipping at shallow angles (only about fifteen degrees) beneath the continent.

The most widely accepted theory for why the Laramide Orogeny produced uplift so far inland in North America is that a segment of the Farallon Plate may have been subducted beneath the continent at a very low angle. Relatively hotter temperatures of this slab segment may have caused this shallow dip angle, or perhaps an oceanic plateau was caught up in the subduction. The latter description could well explain why the Colorado Plateau experienced this uplift without a lot of deformation.

Before the Laramide Orogeny, subduction with a normal dip angle of about thirty degrees produced a line of continental volcanoes where the Sierra Nevada and Peninsular granites are located today. As the slab dip decreased during the Laramide, it caused increased friction on the bottom of the continental plate beneath the Colorado Plateau, wrinkling the crust above. As the Farallon Plate was finally consumed beneath North America, the East Pacific Rise rammed into the side of North America and the San Andreas Fault was born.

It is difficult to determine to what specific elevation the Colorado Plateau may have been uplifted during the Laramide Orogeny, but many geologists surmise that a significant fraction of its total uplift was attained at this time. This is reflected by the amount of rock strata that was stripped off of the southwest edge of the plateau and by some deep paleocanyons that were carved there at this time (see pages 122–123). No matter how high the plateau was uplifted, it remained low compared to the Mogollon Highlands and Rocky Mountains, which towered above it on either side.

The Mid-Cenozoic Orogeny affected the region of the Mogollon Highlands where the Basin and Range Province is located today. As the compressive forces of subduction gave way to the strike-slip forces associated with the San Andreas Fault, the earth's crust became stretched and thinned, causing the Mogollon Highlands to founder. The Mid-Cenozoic Orogeny may or may not have raised the Colorado Plateau (there is no direct evidence for it), but it did begin to change the relative elevation between the plateau surface and areas to the southwest of it. Drainage reversal may be the single largest influence that the Mid-Cenozoic Orogeny had on the plateau surface, and this drainage reversal may have initiated stream piracy toward the north, and/or facilitated the initiation or continued presence of closed, interior basins on the plateau surface.

Late Cenozoic uplift was first inferred from observations on the apparent youthfulness of the plateau's many canyons. A straightforward explanation for this apparent youthfulness was that the region experienced a recent uplift such that the Colorado River and its tributaries kept pace by slicing down into the uplifting terrain. Geophysicists provide more convincing lines of tectonic evidence for this uplift episode (see the discussion of upwelling asthenosphere on page 112), while others advocate for a more passive or isostatic response to drive recent uplift. A combination of the two is also possible. Some of the old questions regarding the timing and mechanisms of plateau uplift are now being answered as more sophisticated tools and techniques are used.

the northeast. Extraordinarily, this is opposite to the flow direction of the Colorado River today. Some aspects of this ancestor to the Colorado River have been known for decades, but more details are beginning to emerge.

The larger geologic setting of the American West about 80 to 30 million years ago reveals that a mountain range, similar in its broad outlines to the modern Andes Mountains in South America, existed to the southwest of the Grand Canyon region. This ancient range of mountains is known as the Mogollon Highlands and existed in a line that connects the modern-day cities of Las Vegas, Nevada; Needles, California; and Prescott, Phoenix, and Tucson, Arizona. It was within this range of mountains that an early ancestor to the Colorado River originated. Many of the newer details about this ancestor of the Colorado River were presented at the 2010 workshop.

Steven Davis, William Dickinson, and others used evidence obtained from tiny zircon crystals plucked from sandstone in the Colton Formation in northern Utah (see sidebar on pages 120–121). They propose that a large trunk river with dimensions comparable to the modern Colorado and Green Rivers flowed north from the Mojave Desert region in southern California to the Uinta Basin in Utah. They named this seven-hundred-mile-long river the California paleoriver, after the source area where it originated (likewise, the modern Colorado River received its name for the same reason). They suggest that this river was positioned in some unspecified portion of the Grand Canyon region, perhaps between the Kaibab and Monument upwarps. Davis, Dickinson, and others clearly did not say that the California paleoriver was involved in any way with the cutting of the Grand Canyon or the placement of the modern Colorado River. They merely reported that the tiny zircons collected in the Uinta Basin were derived from a bedrock source in the Mojave Desert region of southern California (the Mogollon Highlands).

Ivo Lucchitta and Richard Holm resurrected an old idea with new data from an area near the Gap, a prominent break in the walls of the Echo Cliffs east of Grand Canyon. They reported on gravel deposits from Crooked Ridge that trend to the southwest for thirty miles from the Kaibito Plateau toward the Gap. Analysis of the gravel types showed that some of them were derived from as far away as the San Juan Mountains in southwest Colorado. Lucchitta and Holm interpreted this as evidence that a major stream might have existed in the area between about 20 and

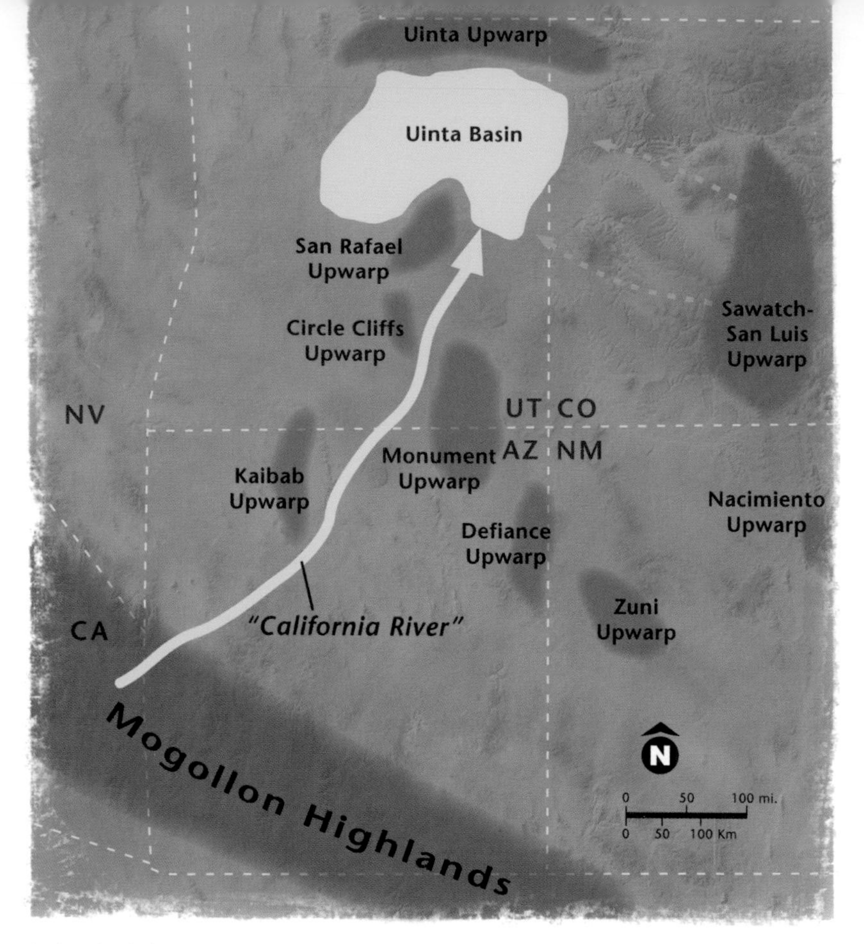

Using detrital zircons, geologists found a new source area for sand in the Colton Formation in northern Utah (the old source area is depicted by the dotted orange lines). The new source area is located in southern California, and a postulated "California River" is what delivered the sediment to the north. Note that this California River is distinct from the one proposed by Wernicke and Flowers.

8 million years ago. They proposed that the Crooked Ridge river drained west into the area of the Grand Canyon, although they stated that it bears no relation to the present drainage configuration.

An Old Grand Canyon

Theories related to an old Grand Canyon are nothing new, but at the 2010 workshop, few geologists were prepared for an innovative idea presented by Caltech professor Brian Wernicke. In collaboration with one of his postdoctoral scholars, Rebecca Flowers, and geochemist Ken Farley, he used a recently developed laboratory technique known as uranium-thorium/helium dating [(U-Th)/He]. When combined with data from a more

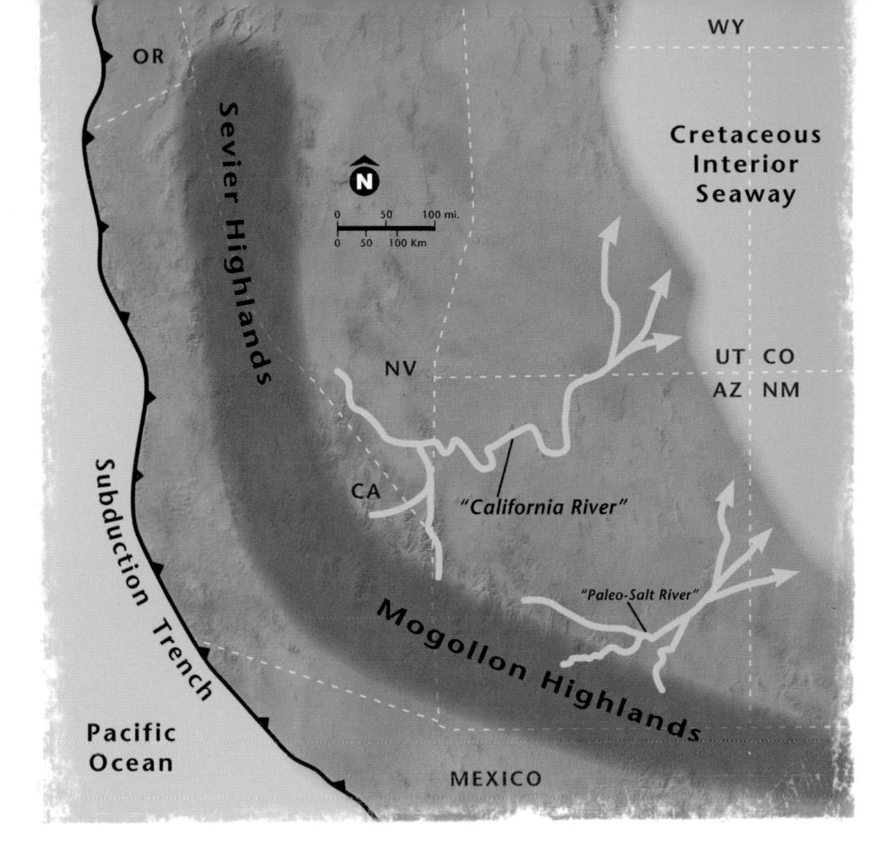

Brian Wernicke and Rebecca Flowers published evidence that suggested how a "California River" might have carved an early incarnation of the Grand Canyon about 70 million years ago. Note that this California River is distinct from the one of Davis, Dickinson, and others.

conventional method (apatite fission track dating, or AFT) used a decade earlier, the data allowed them to determine that a large-scale river might have carved portions of the Grand Canyon about 70 million years ago. The technique revealed the former depth of certain rocks in the modern canyon. Using this method, they propose that western Grand Canyon was cut to within a few hundred meters of its present depth and that eastern Grand Canyon was the site of a canyon of similar proportions to the modern gorge, but cut into higher, now-eroded Mesozoic-age rocks. The findings contradict many long-held observations about the youthful appearance of the canyon, and their theory astounds some geologists. Nonetheless, it is in agreement with other lines of evidence for an old Grand Canyon and uses cutting-edge tools to arrive at the conclusion.

The Caltech group first published their work in 2008, and Wernicke further refined the idea at the 2010 workshop. The technique measures

Use of Detrital Zircon to Identify Ancient Rivers

Like their modern counterparts, ancient rivers contained rocky debris that was derived from the source area. If a river has its headwaters in a highland composed of granite, all of the constituent minerals such as quartz, feldspar, and mica will have been weathered and washed into that drainage system. Some minerals, however, cannot withstand the rigors of river transport, leaving only the most durable grains to survive and be deposited in a downstream basin. This explains why most sand on a beach is composed predominantly of quartz grains—quartz has the capacity to endure the chemical and intense physical weathering involved in river transport. Another durable mineral that can withstand long transport distances is zircon, and it has become a valuable tool in revealing the source area for ancient river deposits.

Zircon is a common mineral found in granite or volcanic rocks, and it endures the rigors of river transport for thousands of miles. When formed in granite, it initially contains a high concentration of uranium but low amounts of lead, which uranium will ultimately decay to. Therefore, measuring the ratio of uranium to lead in detrital zircon crystals will yield the age of crystallization in its formation. This age is usually associated with some known mountain-building event and thus some particular mountain source.

Tiny zircon crystals are extremely durable in river transport and can reveal the source area for a sedimentary deposit. Arizona LaserChron Center-University of Arizona, supported by National Science Foundation NSF-EAR 1032156

The Colton Formation, shown here from the Green River, contains detrital zircon from the Mojave Desert region. Photograph by Wayne Ranney

About nine different mountain-building events have been documented for all of North America in the last 2,000 million years. When the granite that contains zircon is exposed to erosion, the zircon grains travel downstream and are incorporated into a sedimentary rock like sandstone. The age of the zircon reveals what mountain-building event and thus what mountains they originated in. Certain sandstones may contain zircon that was derived from one mountain belt, while others may yield zircons derived from a separate belt. The source area of the deposit is what is revealed.

As they relate to the early history of the Colorado River, detrital zircon studies reveal where certain river deposits originated and what direction the rivers flowed. For example, the Colton Formation found in the Uinta Basin on the northern part of the Colorado Plateau was previously thought to have originated in local sources only: the nearby Laramide uplifts in Colorado and Wyoming. But the detrital zircon evidence showed that the grains originated instead from a source in the Mojave Desert region in southern California, necessitating a complete reevaluation of the origin of the deposit. Studies in the Salton Sea area of California found zircons derived from the Rocky Mountains, therefore revealing when the Colorado River finally became integrated (about 5 million years ago) because of zircons derived from the Rocky Mountains. Studies involving tiny zircons are revolutionizing how geologists understand ancient river deposits, and zircons will continue to be used to better understand the early history of the Colorado River and its ancestors.

the rate of decay of uranium and thorium atoms within the mineral apatite (see sidebar on pages 124–125). The method tells the cooling history of a rock as it is progressively brought closer to the surface by erosion of the overlying material. Their data showed that in eastern Grand Canyon the Coconino Sandstone, the Esplanade Sandstone, and the Vishnu Schist were each buried *under equal thicknesses of rock* during the Laramide. Since these layers are exposed at different elevations within the canyon today, the observed burial depths suggested to them that a similar canyon of roughly the same proportions as the modern one was once present here. Their data from western Grand Canyon is even more startling and indicated that a single pulse of erosion 70 million years ago carved the canyon there. The study further showed that the Paleozoic rocks in eastern Grand Canyon were sliced open and exposed 16 million years ago. These are truly remarkable findings, but some geologists are holding off support of the results until the techniques can be more widely utilized and understood.

Sue Beard and Jim Faulds expanded on previous work by Dick Young and George Billingsley regarding three Laramide-age paleocanyons that have been documented on the Hualapai Plateau on the southwestern edge of the Colorado Plateau. These canyons are the Meriwitica, Milkweed, and Peach Springs paleocanyons found respectively from north to south. Drainage in these paleocanyons came from the Kingman arch, a Laramide-age high that was likely positioned on the eastern edge of the Mogollon Highlands. The canyons were cut through northeast-tilted rocks such that the bedrock floors of their valleys were progressively cut down to younger strata in the northeast direction. The canyons contain deposits bracketed between 55 and 18 million years, with the Peach Springs paleocanyon being the deepest and widest at about four thousand feet deep and three miles wide. Beard and Faulds attempted to document if drainage in these paleocanyons is preserved farther west in mountains in the Basin and Range. Whether these canyons acted as tributary streams to Wernicke's Laramide-age Grand Canyon is unknown, but a connection might be possible. The recognition of these paleocanyons gives geologists a fantastic view of the erosional and depositional history of the southern plateau area during the Laramide Orogeny.

Carol Hill and I published work relating to a possible paleo–Grand Canyon that formed in a northeast-directed river system utilizing other

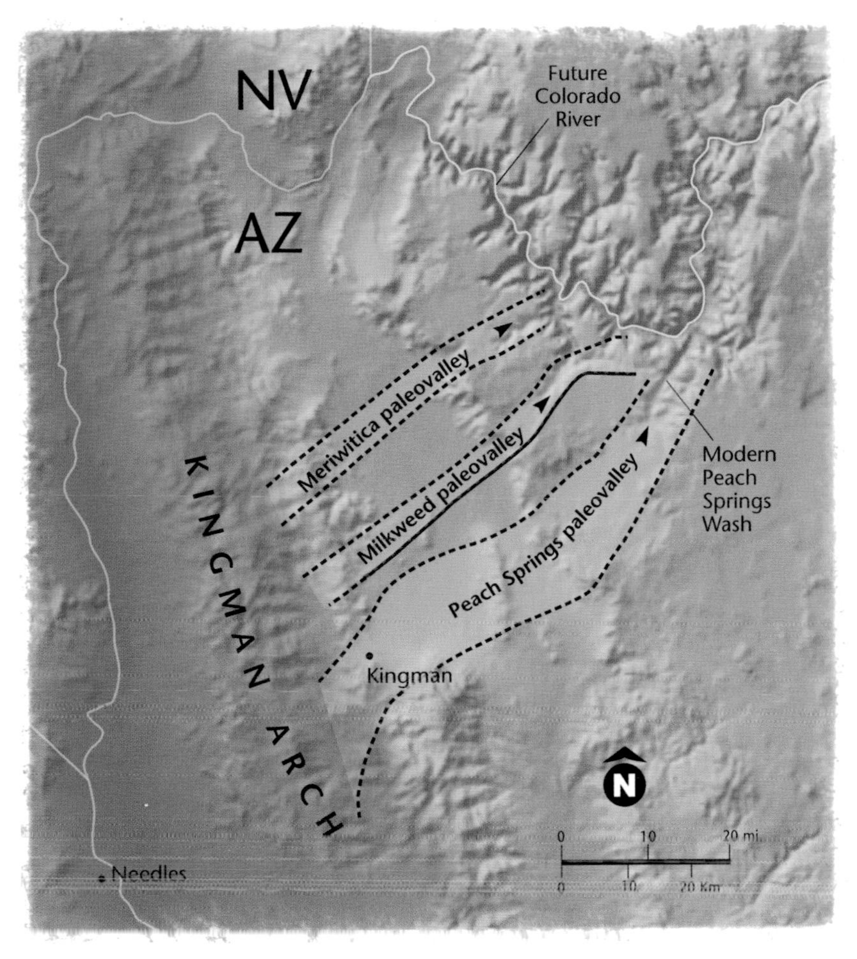

Three Laramide-age paleovalleys have been recognized in northwestern Arizona and may have funneled drainage from the Kingman arch toward Grand Canyon.

well-documented paleo routes on the Hualapai Plateau. Our paleocanyon continued north from Peach Springs Wash along the Hurricane Fault zone. It left the fault zone to the northeast by taking advantage of a recognized joint pattern in the bedrock along the present course of the Colorado River (but going in the other direction). We speculated that the old river might have gone north parallel to the present-day route of Kanab Creek and into the Claron Basin in southwest Utah. This setting is positioned too far west to accommodate the California paleoriver of Davis, Dickinson, and others but is in general agreement with other paleocanyon hypotheses.

Dating Techniques

Geologists use a variety of advanced dating techniques to determine the age of rocks or certain geologic events. Radiometric dating uses the known and constant decay rate of certain radioactive elements in rocks by measuring the ratio of an original parent product to its daughter products. Potassium-argon (K/Ar) dating is one such decay series used with potassium having a half-life of 1,250 million years, meaning that half of the potassium atoms turn into the byproduct argon atoms after that amount of time. After another 1,250 million years, half of what was left will then decay. Some contamination complications once hindered the technique, but these are known and can be corrected for. The argon-argon ($^{40}Ar/^{39}Ar$) method (which looks at the ratio between two species of argon) is increasingly used because it gives more precise dates and suffers from fewer contamination issues. Different decay series yield quite similar results, and the field is considered effective in determining the age of various rock types.

Apatite fission track dating (AFT) and uranium-thorium/helium dating ([U-Th]/He) are valuable and increasingly useful tools to understand when rocks are brought closer to the surface from depth. AFT dating is a process that looks for the microscopic fission tracks that form in apatite crystals when a particle is dislodged from uranium 238 atoms (^{238}U). These fission tracks are unstable at higher temperatures and will disappear when the rock

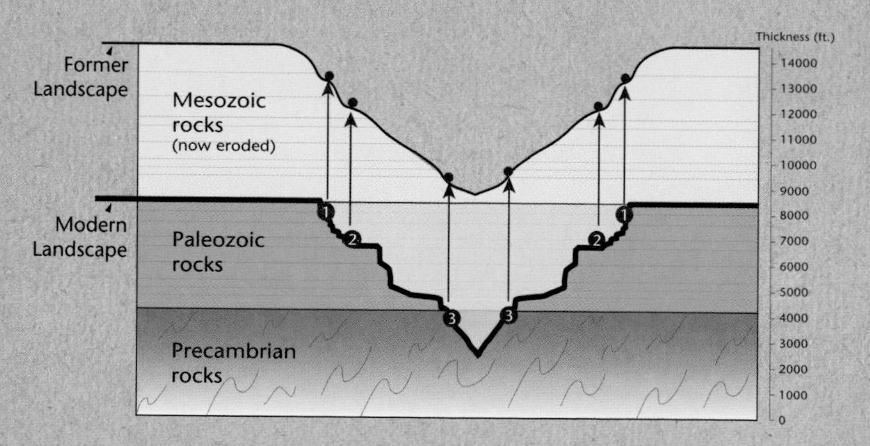

Using apatite fission track dating techniques, some geologists suggest the possible existence of a former canyon that was cut into now-eroded Mesozoic rocks in eastern Grand Canyon. Apatite was collected from the Coconino Sandstone (1), the Esplanade Sandstone (2), and the Vishnu Schist (3). Each apatite sample, although widely separated in elevation today, was buried in similar amounts of rock 70 million years ago. This suggests that a canyon of similar proportions once existed here.

remains at depth. As uplift and erosion bring the rock closer to the surface, the temperature in the crystal decreases and the tracks harden and become preserved within the crystal.

For the related (U-Th)/He dating, the decay of uranium and thorium in apatite will produce the byproduct helium. Because helium is an inert gas, it does not exist as an original, or parent, product in apatite. Thus by comparing the amount of uranium and thorium parent products to the helium byproduct, the time when the apatite crystal began to accumulate helium can be calculated. An age can be determined when the sample cooled below a certain temperature, generally about 230°F (110°C) for fission tracks and about 150°F (65°C) for helium. Combining the two methods—track length and density measurements, and the age of the apatite—will yield information on the sample's thermal history and thus its burial and unroofing history.

Cosmogenic dating is a technique that shows the length of time that a rock surface has been exposed to the atmosphere. It is used to date erosion surfaces, lava flows, rockslides, and many other geologic surfaces that have been exposed for 30 million years or less. The procedure uses the byproducts of cosmic rays in determining when old surfaces were formed. The earth is being bombarded constantly with cosmic rays (composed of highly charged protons and alpha particles), and as these rays strike a rock surface they dislodge certain protons or neutrons to create another element (or a different isotope of the same element). These new products are called cosmogenic nuclides, and scientists measure the concentration of these nuclides to determine how long the sample has been exposed to cosmic rays. Incoming cosmic rays may be affected by elevation, latitude, solar wind, or the earth's magnetic field, but these nuances can be corrected for. The two most frequently measured cosmogenic nuclides are beryllium and aluminum (^{10}Be and ^{26}Al), which are uncommon on newly exposed rock surfaces but form when cosmic rays strike oxygen and silicon atoms (^{16}O and ^{28}Si), two very common original products in rocks. Old surfaces across the Colorado Plateau are now being dated for their exposure ages using cosmogenic dating.

During the last decade, these tools have become useful methods in the study of landscape development (such as canyon cutting), in tectonic geomorphology (how surfaces are raised or lowered), and in dating the age of mountains and sedimentary rock burial. In the Grand Canyon, AFT and (U-Th)/He dating suggest that parts of the canyon may be 70 million years old, with erosion carving western Grand Canyon to near its present depth at 70 million years and eastern Grand Canyon by about 16 million years.

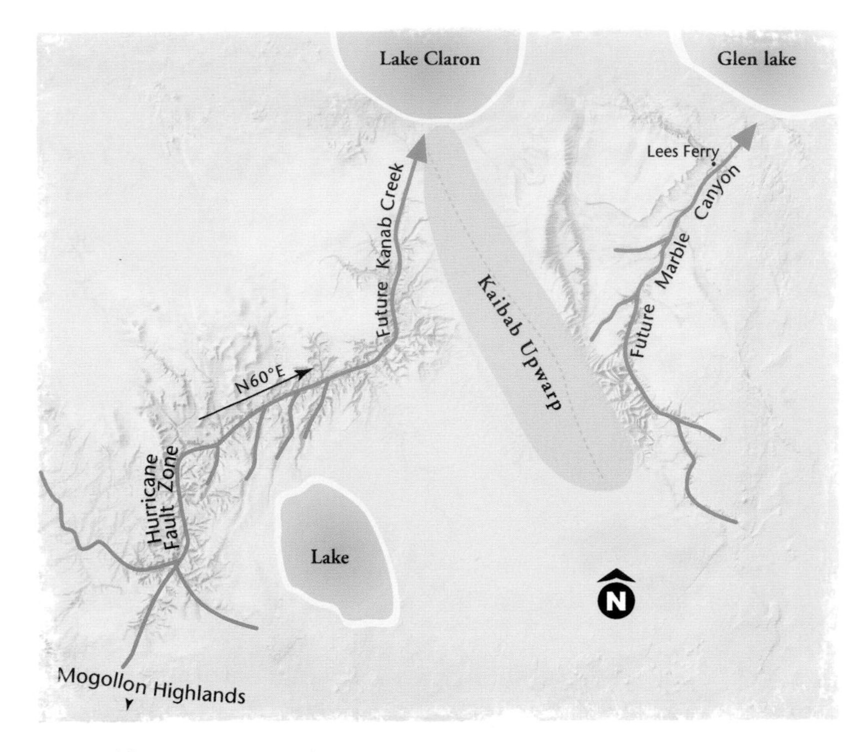

A Laramide-age proto–Grand Canyon as proposed by Carol Hill and me involved two drainages: a western system from the Mogollon Highlands to Lake Claron and an eastern system along the course of the future Marble Canyon to Glen lake.

All of these studies relating to an old canyon emphasize the view that the region was never a featureless plain simply awaiting some future time when the Grand Canyon would begin to appear. Throughout time, at every step, this was an evolving, functioning landscape that ultimately led to the present landscape. It is merely the job of the geologist to piece together how it happened.

Specifically, the studies show that older canyons were cut to significant depths in the immediate Grand Canyon region during the Laramide. In combination with the decades-long research that Dick Young has completed on the paleocanyons and river gravels on the Hualapai Plateau, this new evidence suggests that canyon cutting is not limited exclusively to the most recent period of plateau uplift. The evidence bears out that early uplift of the plateau surface led to significant amounts of dissection near the Grand Canyon. The rub is whether these old canyon systems have a direct link with the modern system of canyons. Virtually no deposits remain that can unequivocally connect the older system of drainage with the present one.

Lack of Evidence for Mid-Cenozoic Drainage across the Region

Northeast-directed flow on the southern Colorado Plateau lasted for 50 million years, from about 80 to 30 million years ago. The rock record then becomes substantially muted for the next 10 to 15 million years. In the Grand Canyon there is virtually no rock record from about 24 to 6 million years ago. This gap in the record may result from the collapse and initial destruction of the Mogollon Highlands during the Mid-Cenozoic Orogeny. Or the dearth of evidence may reflect a gradual change in the climate from the more humid conditions of the early Cenozoic to more arid conditions in the mid-Cenozoic (arid conditions tend to suppress river deposition). Rivers in northern Arizona either lost their highland source area to the south through faulting or suffered diminished precipitation in a global climate shift, or both.

Deposits from this period are scarce, but one suggests that aridity may be the major cause for diminished river flow at this time. The Chuska Sandstone is discreetly tucked beneath the high forests of the Chuska Mountains in northeastern Arizona, with equivalent deposits found sporadically along the Arizona–New Mexico state line and in the subsurface of the Albuquerque Basin in central New Mexico. The thickness of the deposits in the Chuska Mountains are about seventeen hundred feet (slightly less in the other two areas), but the top is eroded, and the deposits may have been somewhat thicker before volcanic rocks covered them about 25 million years ago and aided in their fortuitous preservation. The Chuska Sandstone is dated between about 33 and 25 million years and was deposited in a windblown sandy desert.

Steve Cather estimates that the Chuska "sand sea" may have been one huge dune field measuring approximately fifty-four thousand square miles and nearly half the size of the modern Colorado Plateau (although it likely did not reach as far west as the Grand Canyon area). Cather notes that sand might have been brought initially onto the plateau surface by the northeast-directed rivers that flowed out of the Mogollon Highlands. This drainage possibly became blocked when volcanic eruptions began

Outcrop of the Chuska Sandstone located at 7,000 feet in the Lukachukai (Chuska) Mountains in northeastern Arizona. This may be a remnant of a once extensive "sand sea" on the Colorado Plateau 30 million years ago. Photograph by Wayne Ranney

in the southern Rocky Mountains between 38 and 28 million years ago. Cather indicates that the volcanic highlands could have served as a barrier to the wind, causing the sand to accumulate west of the volcanoes (a smaller version of this setting is found today at Great Sand Dunes National Park near Alamosa, Colorado). Interestingly, ground penetrating radar has detected river systems that are buried beneath windblown sand in the eastern Sahara Desert in Egypt. These rivers once supported African wildlife about four thousand years ago as the nearby rock art attests.

The accumulation of Chuska sand about 33 million years ago is also significant from a global climate perspective—this is the time when the Antarctic glaciation began and signaled a global change to much cooler and drier conditions on Earth (significant evidence for aridity is also preserved today on the Great Plains). The onset of aridity likely caused older deposits to become reworked and winnowed into windblown sand that became trapped against the rising volcanoes to the east. Cather states that the end of Chuska deposition 25 million years ago correlates with other worldwide evidence of more humid conditions.

In the Grand Canyon region, Dick Young describes scant deposits found on the Hualapai Plateau. The rocks are called the Buck and Doe Conglomerate and are interpreted as a fluvial deposit that was derived locally and deposited under arid conditions. It lies above the Music Mountain Formation, which was deposited during the earlier period of northeast drainage. The interpretation is that the Buck and Doe Conglomerate represents a gradual increase in aridity during the mid-Cenozoic (an ash-fall tuff collected near the top of the layer is dated at about 24 million years) as the earlier, through-going rivers lost capacity to move material off of the Hualapai Plateau. Young theorizes that the Buck and Doe Conglomerate overtopped ancient paleocanyon divides and possibly buried the Hualapai Plateau in a few hundred feet of gravel.

Ron Blakey, Travis Loseke, and I reported on fluvial deposits found at the base of the Mogollon Rim, showing that a drainage reversal occurred near Sedona about 26 to 22 million years ago. We showed that northeast-directed drainage from the Prescott area was still present when the Mogollon Rim came into existence about 30 million years ago, but the growth of this escarpment across the regional drainage caused rivers to be deflected toward the southeast near Sedona. The deposit yielding this information is called the Beavertail Butte formation, and I suggest that it likely records the initiation of an ancestral Verde River that went toward the Tonto Basin. These findings show how drainage in the Prescott, Sedona, and Tonto Basin areas might have become partially reversed through the growth of newly developing erosional landforms.

One process that has received sporadic attention from geologists is backtilting of the Colorado Plateau surface. The uplift of the Mogollon Highlands in the early Cenozoic imparted a distinctive northeast tilt on the bedrock strata on the southern Colorado Plateau, and this partially explains why rivers initially flowed that way. Some geologists propose that this tilt might once have been more pronounced than seen today and became partially reversed when the Basin and Range formed, beginning 17 million years ago (rocks from the Mogollon Rim north to Black Mesa still retain a discernable northeast dip). Robert Xavier, Robert Moucha, and others present evidence that backtilting might have been facilitated as upwelling mantle gradually migrated beneath the central part of the plateau surface in the last 30 million years.

The few scattered deposits on the southern Colorado Plateau from the time period 30 and 16 million years ago seem to indicate that rivers became compromised at this time. This is likely related to big changes in the climate or to the topographic relief that was emerging, or both. These new findings help us to understand why so few deposits remain from this mysterious time interval. The arid climate of the mid-Cenozoic diminished river discharge, and an entirely new landscape was forming.

Karst Connection and Groundwater Processes

A new idea relating to the process to integrate the Colorado River was proposed at the 2010 workshop, and it concerns how groundwater flow through the Redwall Limestone may have aided in the development of the Grand Canyon. Groundwater flow through limestone creates karst, defined as landscapes with plentiful caves, sinkholes, and subsurface water channels. If the conditions are right, a chemical reaction between groundwater and limestone dissolves pathways or caves through the rock and vertical conduits or sinkholes may then develop on the surface. These sinkholes serve to channel additional surface water into a karst system, which may then collapse to create or enlarge surface drainage. The idea is not entirely new even as it relates to the Grand Canyon. Charlie Hunt vaguely considered this in his attempt to explain the Muddy Creek problem. But Hunt did not truly grasp the details of karst geology and proposed it merely as an alternative to a "precocious gully," as he called a headward-incising Colorado River.

Current karst connection models, as proposed by Carol Hill, Noel Eberz, and Bob Beucher, suggest that subsurface processes can provide an alternative for how the Colorado River was integrated in the Grand Canyon. The model proposes that a karst aquifer system was channeled in the subsurface beneath the Kaibab upwarp and discharged into a west-flowing ancestor of the lower part of the river. After this subsurface connection was made across either side of the upwarp, karst collapse and fluvial incision created the modern Colorado River and the eastern Grand Canyon sometime after 6 million years ago.

The theory submits that the Little Colorado River previously drained to the north into a proposed "Glen lake" in southeast Utah. A series of sinkholes then developed from Redwall caves near the present-day confluence of the Colorado and Little Colorado Rivers. These sinkholes served to capture surface water and direct it beneath the Kaibab upwarp, discharging into a west-directed drainage. Collapse of the karst system integrated these two drainages, and the modern river was born. The karst connection model is a subterranean-to-surface process and an alternative to headward erosion (a bottom-up process) and basin spillover (a top-down process).

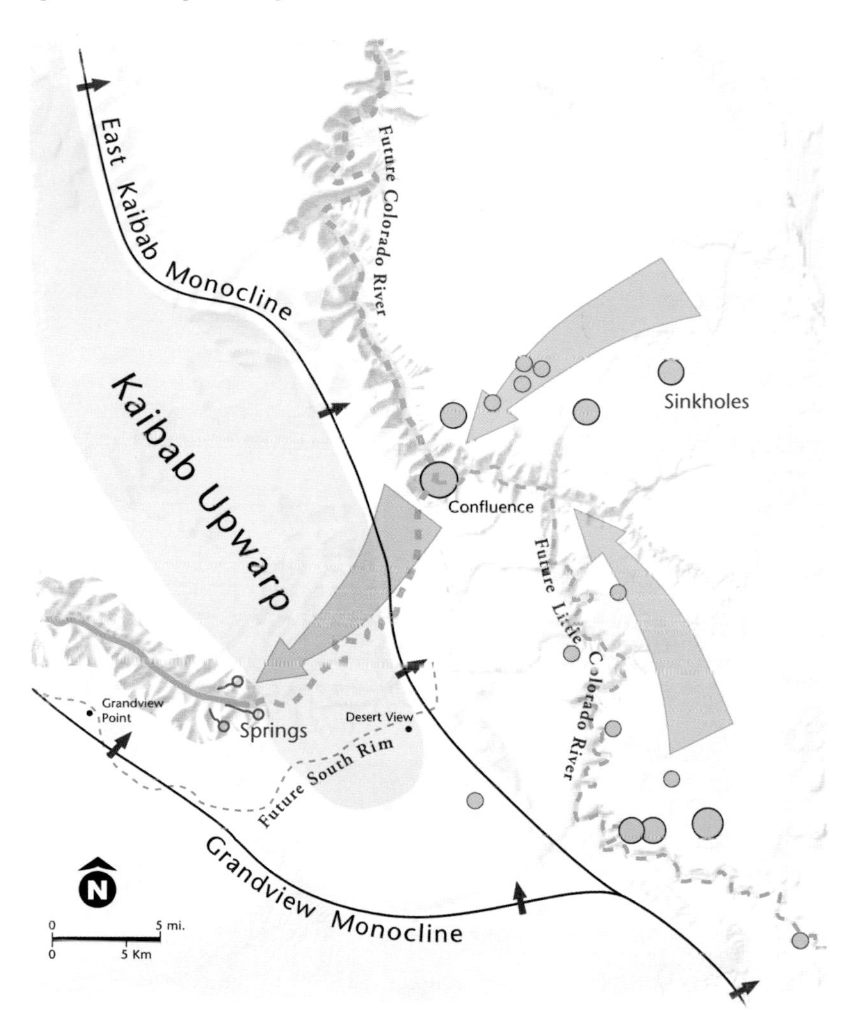

Diagram showing the path of groundwater toward the confluence of the Colorado and Little Colorado Rivers before 6 million years ago. Sinkholes (shown as large blue circles) direct water to the subsurface beneath the Kaibab upwarp and into springs in a headwall beneath Grandview Point. Collapse of this subsurface aquifer may have connected the lower and upper portions of the emerging Colorado River.

View of karst caverns in the Redwall Limestone near river mile 36 in Grand Canyon. Photograph by Wayne Ranney

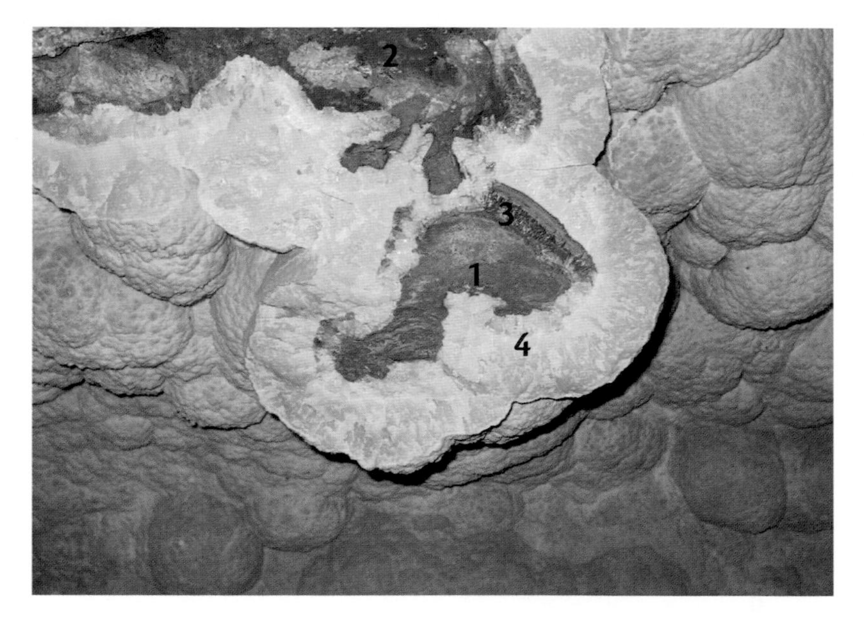

Speleothems form in caves and change as the water table is progressively lowered. Shown here from oldest to youngest are **(1)** bedrock, **(2)** red hematite, **(3)** calcite-spar lining, and **(4)** mammillary coating. The evolving chemical conditions in the cave as the water table is lowered are reflected in these different deposits. Photograph by Victor Polyak

Karst Connections around the World

The word *karst* is derived from the Slovenian word *kras.* The term originated in an area near the border of Italy and Slovenia that includes a barren limestone plateau. The region has numerous limestone deposits with caverns that collapsed, leaving intervening areas with high-standing limestone shapes. The word was first applied to these landscapes in 1894 and is now used worldwide for this type of terrain. Most caverns in the world are the result of karst processes acting upon limestone rock.

Karst theories advanced by Carol Hill and her colleagues do not presently enjoy wide acceptance, but this circumstance may simply reflect an unintended bias by traditional geologists who may be less familiar with karst processes. Many geologists (including me) rarely study the complexities of karst and therefore may not fully comprehend how important karst is in shaping certain landscapes or drainage systems. Hill has worked extensively in karst environments and has found evidence in Grand Canyon for how the karst aquifer system evolved.

Within a cave, various dissolved minerals precipitate gradually on the cave walls to form speleothems (*speleo* = "cave"; *thems* = "deposits"). In the Grand Canyon, speleothems relate to the past level of the water table, and they can be correlated (perhaps) with the incision of the canyon. When a cave is fully inundated with groundwater and well below the level of the water table, a certain type of speleothem called a calcite spar will form on the cave wall. As the water table is lowered, the calcite spar then becomes coated with mammillaries (named for their shape). These mammillaries form at or just below the water table and usually contain enough uranium to be radiometrically dated. As the water table lowers further, gypsum rinds form on those parts of the cave just above the water table. Finally, deposits such as stalactites and stalagmites form after the cave becomes filled with air. Thus, by understanding how different speleothems formed in caves, a record of past water table levels in Grand Canyon can be documented.

Studies like these help determine karst histories around the world and how they have redirected certain drainage systems. Hill points to the Danube River system in Europe, portions of which are now being pirated away from its obvious surface flow toward the Black Sea to the east. Sinkhole piracy here steals water in the subsurface and redirects it to the northwest into the Rhine River system and the North Sea (tracers put in the water system reveal this subterranean piracy). The distance between the mouths of the two rivers is thirteen hundred miles, but the dramatic change in direction occurs in only a seven-mile subsurface reach. Additional support for this theory comes from China, where deep canyons form as the result of karst collapse. In the Chongqing karst of southern China, limestone canyons two thousand feet deep are forming as sinkholes collapse within the karst, integrating surface flow in this area. The karst connection model in Grand Canyon contributes an alternative to a problem so far without a surface solution.

In another study, Carol Hill, Victor Polyak, and Yemane Asmerom worked in western Grand Canyon and obtained uranium-lead dates on distinct cave deposits called speleothems. These speleothems were sampled and dated to reveal that the regional water table was progressively lowered between 17 and 6 million years ago. They interpreted the lowering of the water table as the result of the incision of the western Grand Canyon by headward erosion at this time. Other geologists challenged the interpretation (but validated the usefulness of the method) and suggested instead that the lowering of the water table was better explained by the advent of the Basin and Range. They said that incision of the canyon was not needed to explain the detected drop in the water table. Polyak and others partially agreed but added that the creation of the Basin and Range, with its attendant lowering of base level, is what drove the incision of the canyon. Expect more results from the field of karst studies.

Evidence for a Young Colorado River

Knowing that some geologists favor ideas for an old ancestor to the Colorado River, it might seem exceedingly incongruous for others to present evidence for a quite young river. Yet startling results in this area were forwarded at the 2010 workshop. Working in the area along the lower river near Laughlin, Nevada, and Bullhead City, Arizona (and not coincidentally where Blackwelder first suggested evidence of a young river), Kyle House, Philip Pearthree, and others showed convincingly that closed, disconnected basins became sequentially filled with water, overtopped their bedrock divides, and created a course for the lower Colorado River. Their results show that this fill-and-spill episode spans a 1.5-million-year period that began after 5.6 million years ago and finished by 4.1 million years ago. These studies also helped to clarify the problematic origin of the Bouse Formation.

Their work concluded that four distinct basins contain a similar sequence of deposits that grade from the bottom with (a) material derived only from the enclosing mountains, (b) coarse debris derived from bedrock exposures upstream of the basin edge, (c) fine-grained lake deposits, and (d) unmistakable deposits from the Colorado River. The interpretation is that water rapidly arrived (geologically speaking) in the Las Vegas basin and eventually overtopped a bedrock divide in Black Canyon, thus rapidly

Karst in Chongqing, China, funnels surface water into a subsurface drainage. Photograph by Alexander Klimchouk

filling the basin in modern Cottonwood Valley. The spillover from the Las Vegas basin formed the river through Black Canyon (Hoover Dam area) as it rapidly filled the Cottonwood Valley. Eventually the Cottonwood Basin overtopped a bedrock divide in the Pyramid Hills (Davis Dam area) and spilled water into the Mojave Valley (Laughlin, Bullhead City, Fort Mojave, and Needles areas). This spillover rapidly filled the Mojave Valley, which was large enough to merge with the former lake in Cottonwood Valley, creating a larger and deeper lake. This lake ultimately overtopped another bedrock divide at Topock Gorge, subsequently draining the Mojave Valley and filling the Chemehuevi Valley downstream (Lake Havasu City and Blythe area). This basin was breached by spillover at the Chocolate paleodam, making a final connection with the Gulf of California.

In outlining the sequence of deposits (and the events that created them) the authors provided support for a lacustrine (lake) origin of

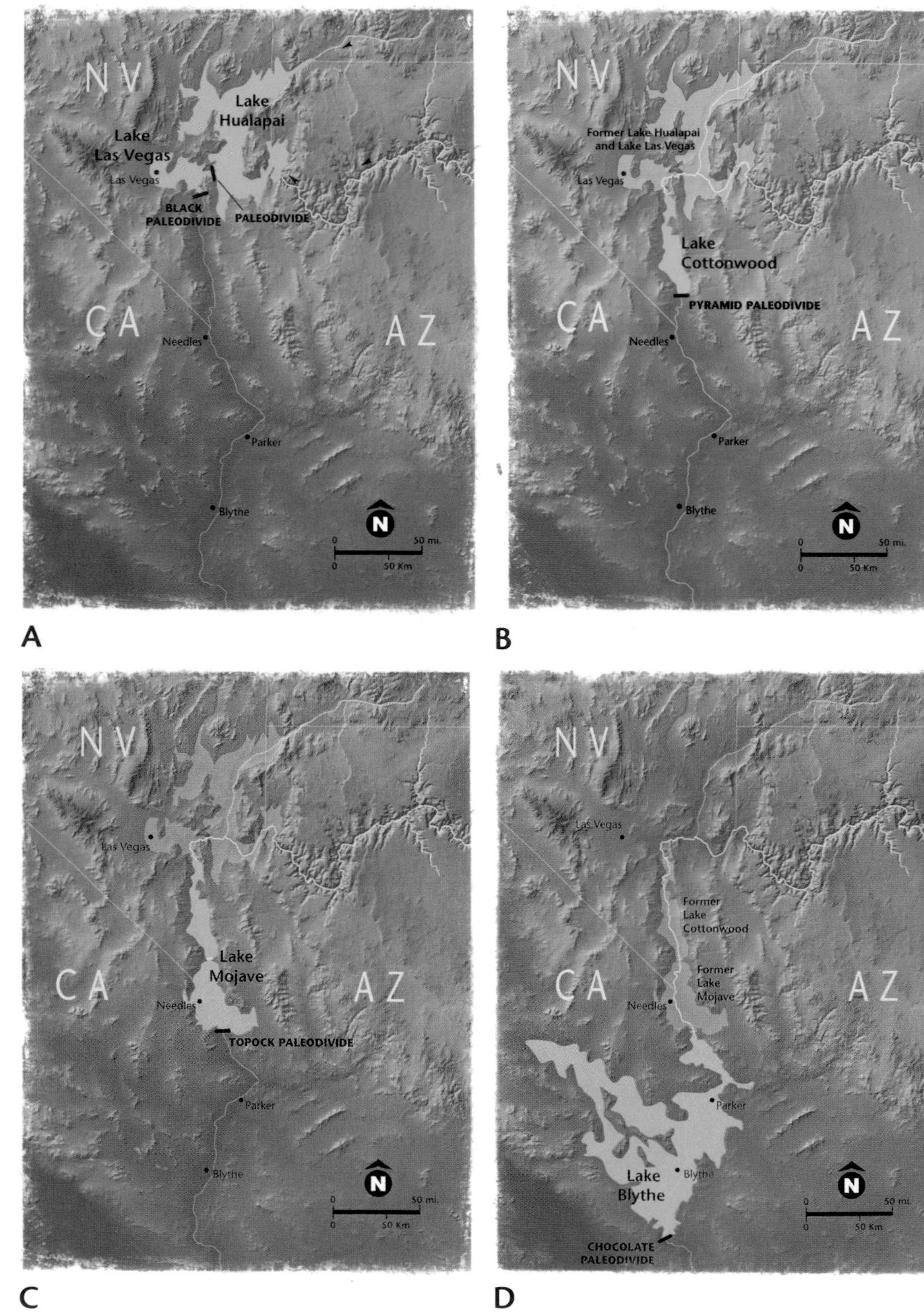
A
NV
Lake Hualapai
Lake Las Vegas
Las Vegas
BLACK PALEODIVIDE
PALEODIVIDE
CA
AZ
Needles
Parker
Blythe
N
0 50 mi.
0 50 Km
B
NV
Former Lake Hualapai and Lake Las Vegas
Las Vegas
Lake Cottonwood
PYRAMID PALEODIVIDE
CA
AZ
Needles
Parker
Blythe
N
0 50 mi.
0 50 Km
C
NV
Las Vegas
Lake Mojave
CA
AZ
Needles
TOPOCK PALEODIVIDE
Parker
Blythe
N
0 50 mi.
0 50 Km
D
NV
Las Vegas
Former Lake Cottonwood
Former Lake Mojave
CA
AZ
Needles
Parker
Lake Blythe
Blythe
CHOCOLATE PALEODIVIDE
N
0 50 mi.
0 50 Km

Shown here is the sequential development of the lower Colorado River from basin spillover.

A Lake Hualapai and Lake Las Vegas form as water reaches these basins. A paleodivide (shown in black) near present-day Hoover Dam confines the basin.

B When the paleodivide is overtopped, the Cottonwood Basin fills, contained by the Pyramid paleodivide.

C Next the Pyramid paleodivide is breached and Lake Mojave forms, drowning the former Lake Cottonwood.

D Finally the Topock paleodivide is eroded and Lake Blythe forms.

E The course of the lower Colorado River is in place after all spillovers are complete.

E

the Bouse Formation, noting only that it was substantially eroded after the lakes drained. Overlying the Bouse remnants are diagnostic and unmistakable Colorado River sand and gravel deposits that culminated in the southernmost basin about 4.1 million years ago, according to them. They wondered what might have brought the rapid arrival of river water to the Las Vegas basin and turned their gaze figuratively and literally upstream toward the Grand Canyon and beyond. Could some other upstream basin also have filled and spilled? Or was it perhaps stream capture related to headward erosion or karst collapse? The results from the lower Colorado River area only intensify questions for how the upper Colorado River became integrated, but the origin of the lower Colorado River now seems rather certain.

Dick Young speculated on a possible setting for the river just upstream from the Grand Wash Cliffs before 6 million years ago. He suggested that a pre–Colorado River canyon might have formed on the Hualapai Plateau by headward erosion into the Grand Wash Cliffs and postulated that the headwaters of this stream were near the Kanab and Cataract drainages. Deposition of the Hualapai Limestone between 11 and 6 million years ago would have been coeval with the development of the precursor canyon, and Young said that as the Hualapai Lake backed up into the precursor canyon, it precluded any coarse sediment from being delivered to the Muddy Creek Formation, thus providing a possible answer to the Muddy Creek problem. This idea picks up on Jim Faulds's ideas from the 2000 conference and coincides with other ideas presented in this chapter.

Jessica Lopez Pearce, Laura Crossey, and Karl Karlstrom studied the Hualapai Limestone near the Grand Wash Cliffs to better understand its origin and depositional environment. They reported that the limestone accumulated between about 12 and 6 million years ago in a series of actively growing basins that slowly evolved from a shallow marsh system to a bona fide freshwater lake. They further suggested that the system was spring fed and sourced by waters chemically equivalent to modern-day Grand Canyon springs (like those in the Havasu Canyon). A major contribution from this study is the interpretation that groundwater sapping of the aquifer system in Grand Canyon (with possible enhanced recharge from the growing San Francisco volcanic field) may have served to integrate the Colorado River and carve the Grand Canyon.

Evidence for what may have happened even farther upstream comes from Andres Aslan and Rex Cole, who studied deposits on Grand Mesa near Grand Junction, Colorado. Their evidence shows that unmistakable Colorado River gravel is preserved by a capping lava flow dated at about 10 million years. They called this river segment the ancestral Colorado River and noted that it was flowing near its present course to the southwest and onto the central Colorado Plateau. The river may have pooled in an interior basin somewhere on the plateau surface, although no deposits of this kind have been found. The river certainly did not flow into the lower Colorado River below Grand Canyon, and the results of their study leave us with a river that is slightly older than 10 million years in Colorado but no older than 6 million years below Grand Canyon. What about the intervening area?

Support for the specific age of the Colorado River east of Grand Canyon so far is inconclusive. Geologists have long proposed that the Bidahochi Formation exposed near Holbrook, Arizona, may document the existence of a large freshwater lake known as Lake Bidahochi (or

Hopi lake) that might have filled and spilled about 6 million years ago, establishing a course for the Colorado River through Grand Canyon. John Douglass presented evidence supporting basin spillover from Lake Bidahochi at the 2010 workshop. He highlighted the criteria that can be used to recognize which of four possible integration processes formed the modern river: antecedence, superposition, piracy, or spillover. Support for a spillover event included: (a) the existence of an upstream paleobasin, (b) the presence of fine-grained (in other words, lacustrine) deposits upstream from a suspected spillover, and (c) the rapid arrival in a downstream basin of water, then sediment. Douglass noted two other criteria that support piracy but found little support for antecedence or superposition.

Jon Spencer and Philip Pearthree also support the idea of spillover from Lake Bidahochi and note that the rate of sediment accumulation before 6 million years ago was only one-third of present-day rates, helping to explain one of the arguments against a large Lake Bidahochi—a lack of thick lake deposits. They assert that when one basin overflowed, it could have rapidly filled the next basin, leaving very little time for significant sediment accumulation. They speculate that Lake Bidahochi could have filled rather rapidly and then spilled to the west and a short-lived Lake Bidahochi would not leave much evidence for its brief existence. There seems to be growing evidence for a young lower Colorado River, but the age for other parts of the system are still being worked out.

The picture that is emerging after two professional conferences in the early part of the twenty-first century is that one of three processes (or some combination thereof) has helped to integrate the Colorado River: headward erosion, basin spillover, or karst collapse. Some elements from the Laramide landscape remain, but the morphology of the Grand Canyon suggests a much later period of downcutting and widening.

The sequence of events that led to the formation of the Grand Canyon still holds some mystery for geologists, but many parts of the story are now well established. These include the initial uplift and resultant northeast drainage in the early Cenozoic, the lowering of base level in the Basin and Range beginning in the mid-Cenozoic, and the role of some process—headward erosion and stream piracy, basin spillover, or karst collapse—in creating the modern Colorado River in the late Cenozoic.

Landscape Evolution of the Grand Canyon Region 5

We have seen that there are too few clues left on the modern landscape that tell us precisely how and when the Colorado River and its Grand Canyon originated. Still, geologists are becoming increasingly familiar with the evidence that does remain—scattered within the canyon, across the plateau, and in adjacent areas. More-modern techniques are employed that enable them to squeeze ever more information from this scant bit of evidence. If we use broad brushstrokes in forwarding an explanation (and understand that some intimate details may never be known to us), we can formulate a general sequence of events that could have given rise to this unique landscape.

John Wesley Powell did not know in 1875, when he proposed that the Green River in Utah was older than the landscape surrounding it, that the modern Colorado River system may be one of the younger landscape features present on the Colorado Plateau. However, the idea of a young Colorado River can be true only if we accept the rather limited definition of the river as a fully integrated, southwest-flowing stream that lies deep within the plateau landscape, a stream for which all or part may be

The Colorado River through Grand Canyon has carved the deepest place on the Colorado Plateau. Photograph by Wayne Ranney

inherited from older and significantly altered river systems. In other words, we can only say how old the Colorado River might be based on how we define what it is and what constitutes its beginning.

A hypothetical example may help. Say that by some stroke of magic, a time-lapse movie of the evolution of the Colorado River was discovered in the historical collections at Grand Canyon National Park. Using this imaginary picture, which would cover the last 80 million years of earth history, we would witness the birth and evolution of the Colorado River and the Grand Canyon. If the movie were four hours long, every second in it would represent 5,555 years of time, while three minutes would document 1 million years of geologic history. If we played the picture in reverse so that the river evolved backward in time to its inception, how much change could we tolerate in the river's configuration, flow direction, or length and still rightly agree that it's the same river? Such a motion picture would begin with a river that is fully recognizable to us, but each frame shown in reverse would change or remove features that might challenge our definition of the river. Some segments of the river would disappear, while others might flow in the opposite direction or become ponded in a lake basin. Could that still be called the Colorado River? Keeping these thoughts in mind might make it easier to more fully appreciate why a precise age for the beginning of Grand Canyon has been so elusive and difficult to determine. Perhaps we can better understand now why some geologists think the Grand Canyon is old while others think the exact same feature is young.

In attempting to unravel the mysteries of the Colorado River and the Grand Canyon, we must be aware that competing and seemingly opposed lines of reasoning can somehow lead to the same landscape result. When anyone favors a particular idea, usually at the expense of another, they are necessarily forced into a much narrower line of reasoning with fewer alternatives possible thereafter. In many ways, the Grand Canyon story is like a maze in that by accepting a viable idea at the start of the discussion, we may be committed to following it through wherever it leads us. Some geologists prefer to emphasize the earliest history of the river and attempt to evolve it forward in time. Others prefer to look at the most recent evidence and attempt to work backward in time. The results can be strikingly different, and an outsider watching it all may wonder if geologists are addressing the same issue. Some of the intimate details of the story remain far beyond our grasp, but in other areas a better picture is clearly emerging.

To help make sense of this complex riddle, it is helpful to subdivide the known history of the Colorado River and the Grand Canyon into discrete events that occurred over time. Six different events are identified, but not each of them is equally represented by the evidence. Some segments of the history still contain unknowns as few (if any) deposits are preserved. However, using what is known to have occurred in adjacent areas and stitching together those parts of the story that have a good record, an outline of events can be formulated:

1. 540 to 80 million years ago: The accumulation of the colorful, flat-lying strata on the Colorado Plateau
2. 80 to 30 million years ago: Laramide-age uplift and initial drainage to the northeast with a possible old Grand Canyon
3. 30 to 16 million years ago: Collapse of the Mogollon Highlands, development of the Chuska sand sea, and unknown drainage
4. 16 to 6 million years ago: Lowering of the Basin and Range relative to the Colorado Plateau with interior or reversed drainage, or both
5. 6 to 4 million years ago: Complete integration of the Colorado River to the Gulf of California
6. 4 million years ago to the present: Recent deepening of the Grand Canyon

A look at each of these will help in understanding how Grand Canyon came to be.

540 TO 80 MILLION YEARS AGO:

Accumulation of Flat-Lying Strata

Prior to the establishment of the Colorado River, the Colorado Plateau was a low-lying depositional basin that accumulated a thick, vast blanket of sedimentary rocks. The future plateau country looked nothing like it does today—the expansive exposure of flat-lying sedimentary rock tells us that the landscape during this time was utterly featureless for hundreds of miles in all directions. It was as unremarkable a landscape as the

A light snowfall accentuates layers of sedimentary rock east of Grand Canyon Village. Photograph by Chuck Lawsen

Mississippi River Delta region is today. More than fifteen thousand feet of sediment accumulated during this 470-million-year time span, most of it deposited at or below sea level. More pointedly, the river environments that once existed here (and represented by the Hermit Formation, for example) were completely buried by younger deposits thus precluding any direct relationship between them and any ancestor to the Colorado River. There is no connection between this vast period of sediment accumulation and the development of the modern river.

80 TO 30 MILLION YEARS AGO:

Laramide Uplift and Initial Northeast Drainage with a Possible Old Grand Canyon

Rivers obviously cannot exist in areas submerged by the sea, and consequently, the story of the Colorado River could not begin until after the ocean withdrew from this area for the final time. The retreat of the Western Interior Seaway exposed a low-lying coastal plain that became the blank canvas upon which the Grand Canyon would later be carved. Nascent river drainages were soon established on this virgin landscape as the uplift of the plateau began, indicating a loosely defined time for the Colorado's birth.

The Tropic or Mancos Shale (the same rock unit but known by different names on sundry parts of the plateau) is the deposit left behind by this final sea. Its nearest exposure to the Grand Canyon today is west of Glen Canyon Dam, but the shale likely once extended much farther south before erosion stripped it back to its present outcrop position. This sea began retreating toward the northeast about 90 million years ago and was gone completely from Arizona and Utah by 70 million years ago. A large mountain range existed to the southwest of the Grand Canyon area, and geologists call it the Mogollon Highlands. These mountains initially formed many millions of years before this time, but their continued uplift expedited the retreat of the seaway. Consequently, the very first drainages to form on this emerging surface flowed from the mountainous southwest toward the coastal plain to the northeast. Ironically, this direction is opposite to the flow of the modern Colorado River.

This initial drainage pattern broadly resembled what can be found today in the northwestern part of South America. Here the continent is colliding with the Nazca Plate and experiencing active uplift that helps form the Andes Mountains. East of the Andes crest, rivers drain in that direction and meander sluggishly onto the floor of the low-lying Amazon Basin, located in modern-day Peru, Bolivia, and Brazil. This setting may serve as a possible modern analogue for how drainage was likely organized on the southern Colorado Plateau between about 80 and 30 million years ago. The Mogollon Highlands formed in a setting somewhat similar to the modern Andes: both were broadly formed in a similar tectonic fashion, and the highlands' former northeast drainage is mimicked by the upper Amazon system. The moist, tropical climate of the Amazon Basin today might even mimic the early plateau climate, which brought about Dutton's Great Denudation and Davis's plateau cycle.

Rocks found on the Hualapai Plateau in Milkweed Canyon (a side canyon within Grand Canyon) further aid the regional tectonic evidence

Between about 95 and 70 million years ago, the Mancos Sea covered much of the future Colorado Plateau. At this time rivers flowed from the Mogollon Highlands to the northeast toward a retreating shoreline. This flow direction is opposite that of the modern Colorado River today. Illustration by Ron Blakey

for northeast drainage. These deposits are known as the Music Mountain Formation, which contains well-rounded rock types that could only have been derived in streams flowing out of the southwest—the only direction where rocks of this kind were exposed at that time. The Music Mountain Formation belongs to a larger group of fluvial sediments that are historically and informally known as the Rim gravel. Together with the known tectonic setting of the region, these deposits confirm that

Significant dissection of Grand Canyon strata during the Laramide Orogeny is documented by the presence of Laramide-age gravel remnants (orange deposit, center) within Milkweed Canyon. Photograph by Wayne Ranney

northeast drainage was present across the Grand Canyon region at this time. Some geologists postulate that the Rim gravel may have completely blanketed the Grand Canyon area and that the eventual course of the Colorado River might have been set upon the swales of this surface. Others aren't so sure.

A larger picture that is well established is that as drainage was directed to the northeast across the Grand Canyon region, it flowed initially to the shore of the retreating seaway, then across the continent to the Mississippi River system. Only at about 60 million years ago, when the Rocky Mountains were uplifted, were the northeast-directed rivers blocked to create freshwater basins on the future Colorado Plateau surface. Partial evidence for these early lakes is preserved in the delicate hoodoos of the Claron Formation in Bryce Canyon National Park.

Overwhelming tectonic evidence and a smattering of river deposits provide an emerging picture of this first chapter in the history of the Grand Canyon. The evidence for Laramide-age paleocanyons in the area is convincing, and some geologists used it to invoke a paleo–Grand Canyon. Others think that early incision is not necessarily related to the modern canyon. Nevertheless, some ancestor to the Colorado River was in existence by 80 or 70 million years ago and remained until about 30 million years ago.

By 65 million years ago, rivers drained across the future Colorado Plateau to the Mississippi drainage in the center of the continent. Illustration by Ron Blakey

30 TO 16 MILLION YEARS AGO:

Collapse of the Mogollon Highlands, Development of the Chuska Sand Sea, and Unknown Drainage

Although it may be difficult to determine if the initial northeast drainage system had any direct influence on the modern placement of the Colorado River in Grand Canyon, this early system of drainage was a relatively simple arrangement that went from the Mogollon Highlands in the south to the future plateau surface in the north. During the next episode, this drainage system was disrupted when crustal disturbances in central Arizona began to affect the highlands to the south of the Grand Canyon area.

This disturbance, known as the Mid-Cenozoic Orogeny, was caused by stresses within the earth's crust that caused it to become stretched and thinned. The highlands foundered down from their once lofty elevation, and this certainly exerted an effect on the streams flowing out of them. Tectonic lowering may have formed catchment basins either at the base of the mountain front or on the Colorado Plateau surface as the free flow of water to the northeast was disrupted. As the mountains were lowered, the climate also changed to more arid conditions. There are practically no deposits in the Grand Canyon area from this period and only a handful on the plateau and in adjacent areas; therefore, it is difficult to know precisely what rivers were doing at this time.

In the Grand Canyon area, a single gravel formation represents this entire 14-million-year time period. It is located on the Hualapai Plateau in Milkweed and Peach Springs Canyons and is called the Buck and Doe Conglomerate. It overlies the older Music Mountain Formation and has some significant differences from it. The clasts in this rock are derived in large part from rocks that make up the walls of the paleocanyons that enclose it. Some older crystalline clasts from the south are also found in this formation, and the combination of these rock types suggests that the Buck and Doe Conglomerate originated in localized streams that came from the south. More important, the deposit is less chemically weathered than the earlier gravels, documenting that a drier climate was present here in the mid-Cenozoic. An ash bed near the top of the deposit is dated at about 24 million years.

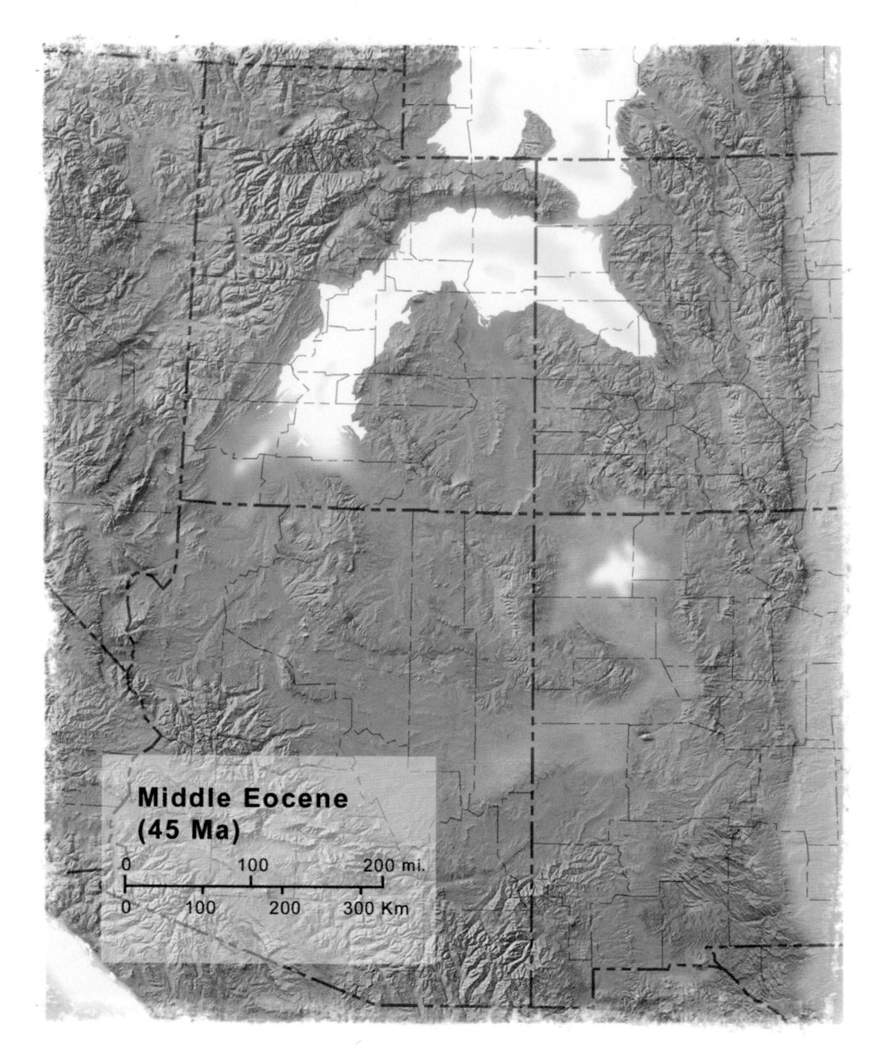

By about 45 million years ago, the Rocky Mountains had been uplifted and interrupted the free flow of drainage such that huge freshwater lakes were developed on top of the future Colorado Plateau surface. Illustration by Ron Blakey

On the plateau in northeast Arizona, another deposit called the Chuska Sandstone represents the remains of a large dune field that might once have covered the entire

Strata in Bryce Canyon known as the Claron Formation represent the former location of one of these lakes. Photograph by Bob and Suzanne Clemenz

Between about 30 and 25 million years ago, the Chuska sand sea was present over much of the southern Colorado Plateau surface. Near Sedona, Arizona, river gravels document that the northeast-flowing streams were interrupted by the growth and development of the Mogollon Rim. Illustration by Ron Blakey

southern Colorado Plateau between about 33 and 25 million years ago. Such a postulated sand sea would have buried and defused any preexisting river system that was present in this area. It is unknown if this is when the northeast system of drainage was destroyed completely or otherwise ceased to have a direct connection with the modern river system. Conversely, the ancient river courses may have been exhumed from beneath their sandy veneer when sand accumulation ceased. It is difficult to imagine this latter scenario, but the older channels could have reappeared with perhaps reversed flow direction.

Gravels from near Sedona, Arizona, document how the northeast drainage was interrupted here by about 25 million years ago. At this time streams were still flowing from the Prescott region toward the northeast but became deflected to the southeast at the base of the Mogollon Rim just south of Sedona. The Rim was formed by erosion in a strike valley that formed parallel to the Mogollon Highlands (strike valleys form when alternating hard and soft strata are tilted, causing the soft layer to erode longitudinally to create a valley between two resistant ridges of harder strata). The deflected streams went to the southeast along the base of the Rim toward an unknown destination southeast of the modern Verde Valley. Perhaps they ponded in the emerging Tonto Basin near present-day Roosevelt Lake.

Since few relationships can be recognized between these scattered deposits, it is difficult to piece together what the larger drainage pattern may have looked like. The arid conditions that developed at this time greatly lessened the discharge of rivers and perhaps flow was seasonal or even nonexistent at times. Ponding within interior basins is a likely possibility as well. Perhaps the paucity of sediments from this time represents the hazy beginning of a major drainage reversal in the region. It seems very probable that any drainage in the Grand Canyon area at this time was "confused," meaning it was dry, interrupted, buried, ponded, reversed, or possibly some of each at various locations. Much conjecture surrounds this enigmatic period because of a lack of good evidence.

16 TO 6 MILLION YEARS AGO:

Lowering of the Basin and Range Province with Interior and/or Reversed Drainage

As the Farallon Plate was finally consumed beneath the western edge of North America, the compression that had actively kept the Mogollon Highlands elevated was terminated. As the oceanic spreading center was subducted, it brought North America into contact with the Pacific Plate, and the relative motion changed from forces of compression to those of extension. The earth's crust now began to stretch and thin where the highlands used to be, and the landscape was significantly modified southwest of the Grand

Canyon area. The Colorado Plateau became a topographic high relative to the newly emerging Basin and Range—in fact, the event is known as the Basin and Range Disturbance. Although it occurred off of the plateau edge, the disturbance played a huge role in the development of the Grand Canyon since future drainages were shaped as a response.

As the Basin and Range was formed and the Colorado Plateau became elevated relative to it, the plateau itself may not have been uplifted. This is an important distinction to note, since the differential lowering of the Basin and Range might be what earlier geologists perceived as a period of plateau uplift. Recall that some geologists once stated, "No uplift, no canyons," but that saying has now been refined to "No lowering of base level, no canyons." The development of the Basin and Range precisely where the Mogollon Highlands once stood could be the most significant event to occur in creating the modern drainage—a dramatic reversal of the drainage system was likely well under way by this time.

Important deposits are located on either side of the Grand Canyon. On the west is the Muddy Creek Formation, laid down in a basin (the Grand Wash trough) that formed when the Grand Wash Fault slipped to define the western edge of the Colorado Plateau. East of the Grand Canyon astride the Little Colorado River valley near Holbrook is the Bidahochi Formation, which did not accumulate in a fault-bounded basin. Some aspects of the formation's history remain unresolved, but its relevance to the Colorado River story cannot be overstated. Both of these interior basins were likely closed, meaning that they had no outlet to the sea. Understanding the evolution of these two basins helps to shed light on the Grand Canyon's origin.

The Muddy Creek Formation, which you will recall contains no recognizable deposits from the Colorado River, lies across the river's path and the interpretation is that the modern river could not have been in existence while the deposit was accumulating (the youngest part of the deposit, the Hualapai Limestone, is between 11 and 6 million years old). Various explanations have been forwarded to explain why this deposit lacks any obvious evidence for the existence of the Colorado River and the Grand Canyon. One possibility is that Hualapai Lake extended up into a proto-western Grand Canyon, thus precluding any obviously visible evidence for the existence of the Colorado River during Muddy

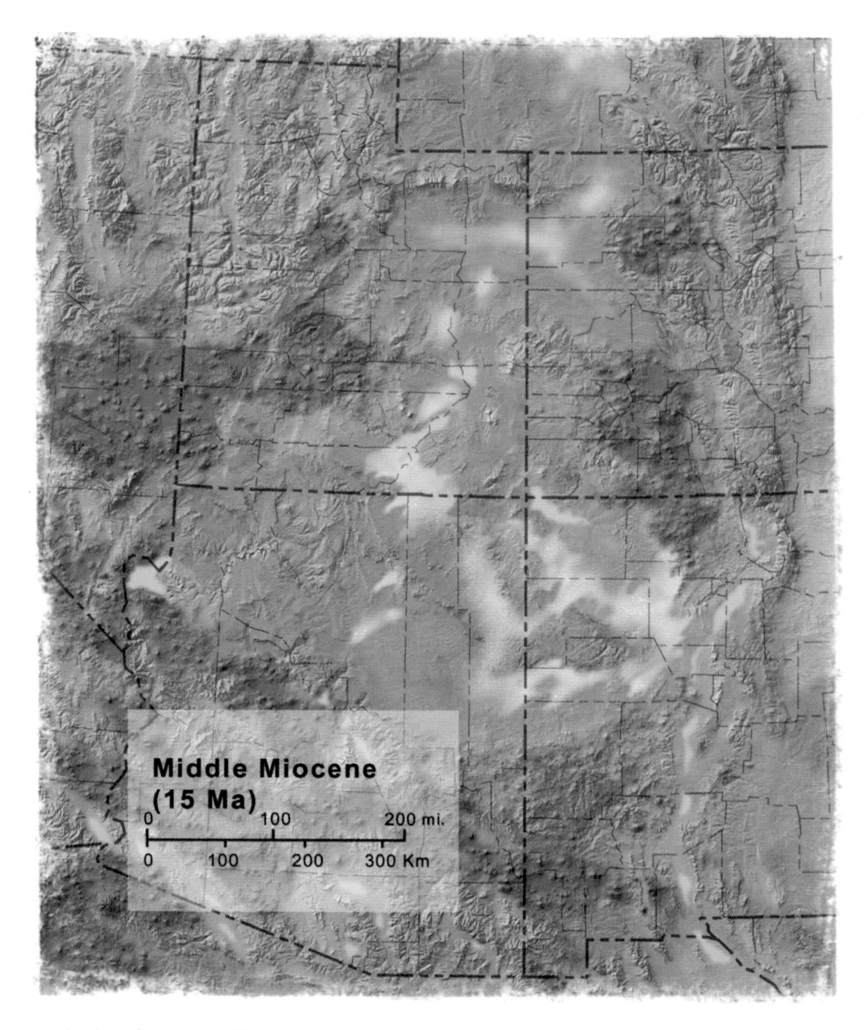

Beginning about 16 million years ago, the Basin and Range Disturbance destroyed the Mogollon Highlands, lowering the landscape west of Grand Canyon and inverting the drainage pattern in the region. Headward erosion into the western plateau edge may have commenced at this time. Both the Muddy Creek Formation and the Bidahochi Formation were deposited between 16 and 6 million years ago. Illustration by Ron Blakey

Creek time. This interpretation is tentative, however, and the simplest explanation may be that the modern river did not exist until after Hualapai Lake drained.

The Bidahochi basin at the opposite end of the canyon contains deposits that are the same age as the Muddy Creek Formation—16 to 6 million years. However, these deposits are equivocal, and some geologists are uncomfortable with the interpretation for a lake of considerable size or duration of time. Some of them speculate that the Bidahochi Formation may have accumulated only locally on an alluvial plain that held small, localized ponds. Others assert that this lake existed as recently as 6 million years ago and then spilled to the west through the Grand Canyon. Both the Grand Wash trough and the Bidahochi basin ceased to receive sediment by about 6 million years ago, and these identical dates from widely separated basins hint that an event of major proportions likely occurred at this time. What could this event have been?

Historically, geologists have speculated that headward erosion into the plateau edge from the Basin and Range by a young, steep-gradient stream could have formed the Grand Canyon. The tectonic lowering of the Basin and Range (or relative uplift of the plateau, if you prefer) would have created the steep gradient necessary for a stream to expand its drainage area in capturing the upper Colorado River. This stream capture event, variably on either side of the Kaibab upwarp, would have worked to destroy two separate interior basins and create the modern Colorado River (and perhaps the Grand Canyon) in the process. It could also explain why the Muddy Creek and Bidahochi Formations ceased being deposited at precisely 6 million years ago.

Hualapai Limestone comprises the upper part of the Muddy Creek Formation and is exposed west of the Grand Wash Cliffs. It is as young as 6 million years and is the last deposit to pre-date the modern Colorado River. Photograph by Wayne Ranney

Other geologists have voiced concern about the efficiency of headward erosion in arid land settings and look instead to the rapid spillover of Lake Bidahochi as the driving integration process. Perhaps a karst aquifer system that had developed in caves in the Redwall Limestone collapsed to establish the course of the river across the Kaibab upwarp. One of three processes—headward erosion, closed basin spillover, or karst collapse—likely served to integrate the Colorado River somewhere near the Kaibab upwarp, perhaps using portions of the preexisting landscapes that were formed during previous episodes of erosion. Another viable consideration is that it was some combination of these three.

The interior basin deposits on either end of the Grand Canyon are all we have to determine the evolution of the Colorado River between 16 and 6 million years ago. They provide some clues to what landscape elements were in existence near the canyon or what elements were not in existence (such as the modern, fully integrated river). Were some segments of the modern river present but still disconnected? How much of the Grand Canyon existed at this time, either in its length or depth? The next chapter is better known.

6 TO 4 MILLION YEARS AGO:

Complete Integration of the Colorado River to the Gulf of California

Good evidence for the development of the modern Colorado River—one that exited the Grand Wash Cliffs and began flowing southwest toward the Gulf of California—appears on the landscape shortly after 6 million years ago and culminates no later than about 4 million years ago. The evidence is represented by the presence of unambiguous Colorado River sediments in the Lake Mead area, other sedimentary deposits along the lower Colorado River corridor, and still more deposits in the area of the Salton Sea. All of these contain sediment types and reworked fossils that are definitely attributed to the modern river, and the evidence has become incontrovertible in recent years.

The first of these definitive Colorado River deposits is a cobble and gravel accumulation that is found near Sandy Point, along the shores of upper Lake Mead. This area is downstream from the Grand Wash Cliffs and shows that by this time the river was exiting the Grand Wash Cliffs. The gravel is capped by a basalt lava flow that has been dated at 4.4

million years. This means that the deposit, and thus the modern river, was flowing past this spot sometime before this date.

A bit farther downstream lie the Las Vegas, Cottonwood, Mojave, and Chemehuevi Basins. The rocks in each basin show a similar sequence of strata grading upward from (a) closed-basin alluvial fan deposits, (b) fluvial boulder conglomerate, (c) lacustrine beds belonging to the Bouse Formation, and (d) fluvial quartz-rich sand, mud, and gravel. The lowermost alluvial fan deposits that lie on the floor of each basin document the pre-integration phase of the system. These rocks are generally overlain by a boulder conglomerate, which has the same composition as the bedrock divides that once defined the upstream side of the formerly closed basins: the interpretation is that the conglomerate is the debris eroded from the divides as they were overtopped with lake water. The overlying Bouse Formation records when each basin was filled with lake water. The uppermost and last deposits of quartz-rich sand and mud have distinctive Colorado River signatures in them. The entire sequence contains datable ash beds, which reveal that the fill-and-spill events eventually cascaded downstream from the Las Vegas basin to the Gulf of California by about 4.8 million years ago. The interpretation is that formerly disconnected closed basins were rapidly filled with river water, which spilled over their lowest rims to hydrologically connect the four basins. The cause of such a rapid delivery of river water to the basins is unknown, but any of the three processes proposed for the integration of the river could have accomplished it.

The opening of the Gulf of California is one of the biggest events in the history of the Colorado River and the Grand Canyon. When the San Andreas Fault came into existence about 12 million years ago, it ripped Baja California apart from mainland Mexico. The upper Gulf of California was flooded with seawater about 7 to 6 million years ago. This formed a new base level where deposits accumulated to more than eighteen thousand feet thick in the Salton Sea area (located about one hundred miles farther south before the San Andreas Fault transported the deposits to their present position). The oldest of these deposits is known as the Split Mountain Formation and is located in the hills west of the Salton Sea, or is known from subsurface drilling. It contains fossils of foraminifers and plankton that document an exclusively marine origin.

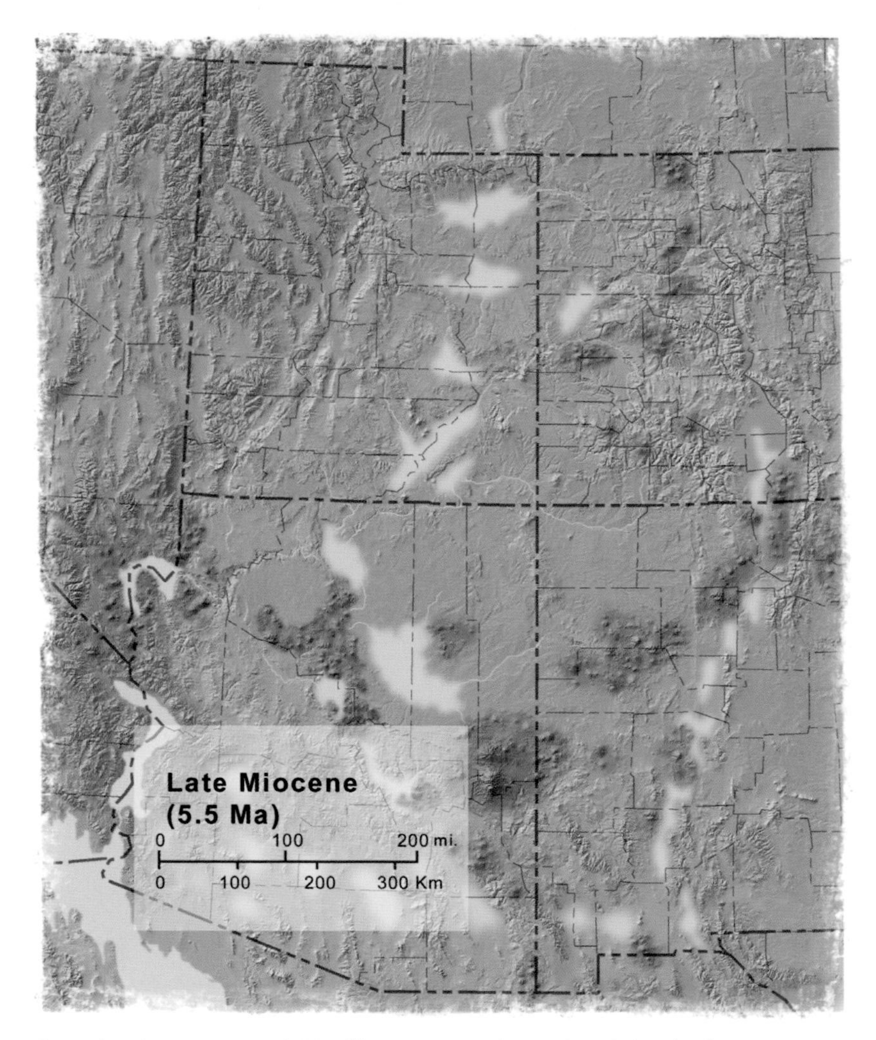

Sometime between 5.8 and 4.4 million years ago, the modern Colorado River was integrated through Grand Canyon to the Gulf of California. Recent interpretations reveal that closed basins along much of the river may have spilled over to create a path for the modern river. Note that this map favors basin spillover above other equally possible processes. Illustration by Ron Blakey

Overlying the Split Mountain marine sediments is the Imperial Formation, displaying an upward progression of marine mudstone near the base of the deposit to marine sediments that show sandy inputs derived from the Colorado and Gila Rivers. This change in sediment type, in conjunction with detrital zircon studies, is well constrained between about 5.3 (Gila River) and 4.8 (Colorado River) million years ago. It is one of the strongest lines of evidence showing when Colorado River water first arrived here from the north. Within the Imperial Formation are small fossils that originated in the Mancos Shale on the central Colorado Plateau. The interpretation is that these fossils were eroded and transported into the Imperial Formation through the Grand Canyon by an integrated Colorado River. Below a specific horizon, these reworked fossils are not found; above that horizon, they are present. Reading the evidence suggests that this horizon documents the moment in time when the Colorado River became fully integrated by some manner into the system we see today.

Recognition of an integrated, through-going Colorado River around 5 million years ago means the river must necessarily be the end result of some longer-lived landscape-forming processes. We are again prompted to ask, Could the Grand Canyon have existed in some formative stage before this time? From the evidence, it seems quite unlikely that the canyon appeared instantly on the landscape just as the river became fully integrated. Some geologists perhaps conflate the specific age of the modern river as a proxy for the age of the canyon. Others look to basic evolutionary concepts that show how previous ancestors have direct links to their descendants. Increasingly, some geologists recognize that parts of the canyon must have existed in some form before the river was fully integrated. Others insist it is a strictly young feature with its youthful morphology perhaps the best evidence for this.

4 MILLION YEARS AGO TO THE PRESENT:
Recent Deepening of the Grand Canyon

Ever since the Colorado River became integrated around 5 million years ago, it has been actively deepening and widening the great gorge. It is common to think of the Grand Canyon as forming in a slow and steady manner, inch by rocky inch. But some evidence suggests that it is deepened in starts and stops, reacting to the multiple inputs of uplift, base level lowering, climate change, and variable discharge amounts of the river. Perhaps as much as one-third to one-half of the current depth of the canyon was achieved in just the last few million years—an amazing thought considering the canyon's huge proportions.

The last 2 million years of earth history are known as the Quaternary Period, and this period is notable because of the drastic climate changes

The Imperial Formation in Anza-Borrego Desert State Park, California, holds evidence for when Colorado River water first entered the Gulf of California. Photograph by Rebecca Dorsey

that plunged the earth into the ice ages. The Rocky Mountains were greatly shaped by the huge glaciers that formed there, but these conditions were not static, and many pulses of glacial advance and retreat are documented. This means that as ice occasionally melted in the Rockies it was channeled down the Colorado River through the Grand Canyon. This increased runoff must have generated megafloods that would dwarf any discharge in recorded history. (The largest flood recorded in the Grand Canyon occurred in 1921, when water at the rate of 220,000 cubic feet per second raged through the gorge during the late spring melt.) Geologists estimate that floods on the order of 1 million cubic feet per second or more—the average flow of the Mississippi River today—may have raged down the Colorado during a glacial retreat. Imagine the rocks and debris carried in such a flood and the power that deepened the Grand Canyon.

Other processes regarding the deepening of Grand Canyon have also been proposed. In western Grand Canyon, the Toroweap and Hurricane Faults cross the river and have actively lowered its bed by at least nineteen hundred feet to the west. The lowering of base level may be an important process in deepening the upstream portions of the canyon. As earthquakes rattled the land, they raised the river's channel upstream, creating a knickpoint obstruction across it. These knickpoints were subsequently attacked by the erosive power of the Colorado River, causing them to migrate upstream and deepen the upstream channel of the river in the process. Together with increased runoff in the river during the Quaternary, knickpoint migration upstream from active fault lines most likely caused eastern Grand Canyon to become significantly deepened at this time. The Upper and Middle Granite Gorges and Marble Canyon may have been deepened in this process in only the last 1 million years. The three canyons are essentially large-scale slot canyons that formed rather quickly in response to increased runoff, the lowering of base level, and subsequent knickpoint migration.

As a river channel deepens, widening of the canyon responds in kind. Alternating hard and soft layers of strata are found within the walls of Grand Canyon, and as the softer strata become exposed, they begin to erode relatively rapidly away from the river's edge. This undercuts the harder layers above them that otherwise would be resistant to the forces of erosion (this is how the Tonto and Esplanade Platforms have developed). Thus, increased deepening in the canyon causes its profile to become over-steepened, leading to mass wasting events (movement of huge masses of earth) that eventually migrate upslope in the canyon. There are a few places in Grand Canyon, such as Hermit Canyon, where this upslope migration is evident as it undercuts harder layers. Numerous megalandslides are also evident within the canyon and document how widening proceeds on a grand scale.

The last seven hundred thousand years hold even more surprises concerning the canyon's spectacular evolution. Between the Toroweap and Hurricane Faults in western Grand Canyon, spectacular lava flows are

Headward erosion is actively attacking the headwall in Hermit Canyon and is visible as the two shadowed indentations cut back into the sunlit, smooth-sloped upper rim of the canyon. Photograph by Wayne Ranney

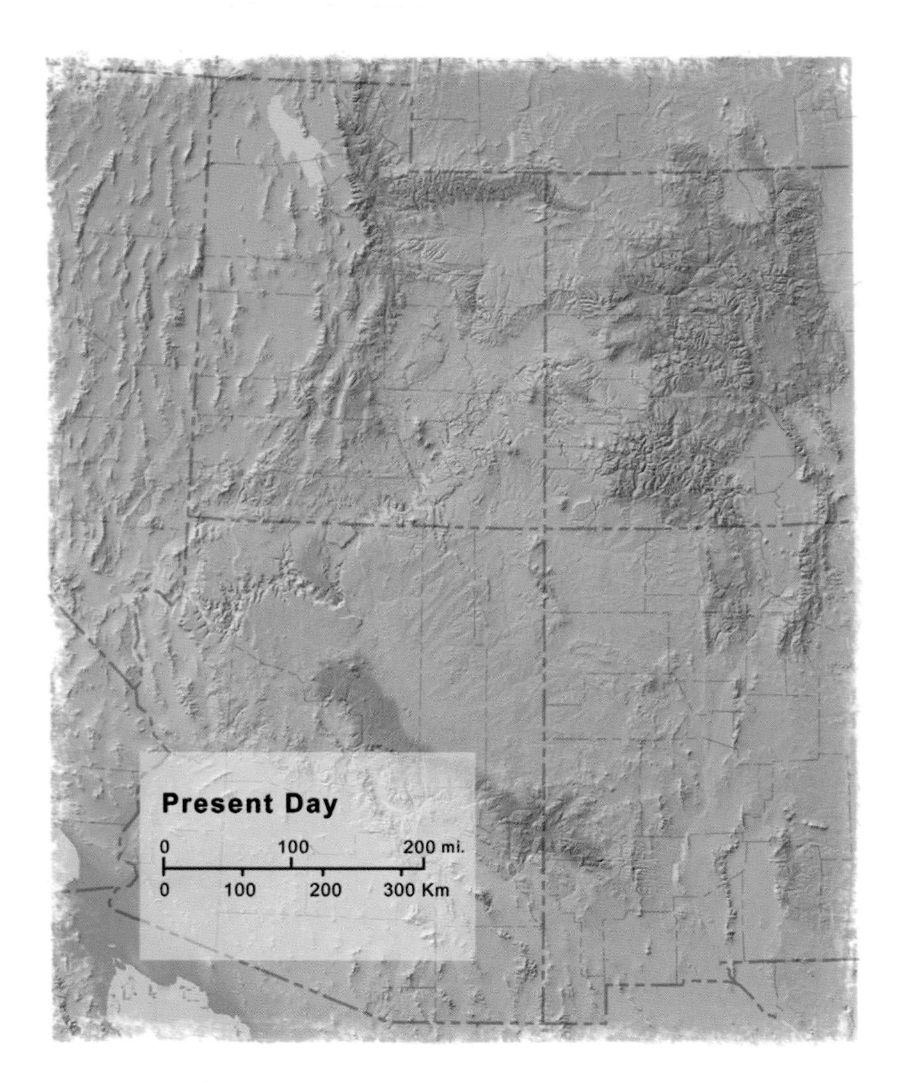

Present-day configuration of the Colorado River on the Colorado Plateau. During the last 6 million years, substantial deepening of the canyon has occurred by one or more processes: headward erosion, basin spillover, or karst collapse. Illustration by Ron Blakey

seen frozen against the canyon walls. Numerous side canyons are observed filled with the outpourings of this basaltic lava rock. The lava erupted so voluminously that huge dams were created across the Colorado River's path in Grand Canyon. One of these dams was more than twenty-three hundred feet high. A fluid lava flow choked the river channel for eighty-four miles. In all, at least thirteen lava dams once blocked the Colorado within the walls of Grand Canyon.

As spectacular as it must have been to witness the red-hot lava pouring into the canyon (and its glacially fed river), it is the geologic lesson learned from this lava story that is truly astounding. A twenty-three-hundred-foot-high lava dam, if it survived quick destruction or collapse, would have created a naturally formed reservoir on the river that would have stretched upstream to an area just below Moab, Utah, over three hundred miles in length. Below Grand Canyon Village, the entire Tonto Platform would have been beneath three hundred feet of water, with Indian Garden Campground two hundred feet beneath the still waters. What a scene this could have been.

Unfortunately, no reservoir deposits have been recognized, and some geologists urge caution when invoking the presence of long-lived lava dams in the canyon. They point out that any lava dams may have been inherently unstable, as they were built on top of unconsolidated river gravel, and that the lava likely shattered as it came in contact with the river. This suggests that some lava dams may have been short-lived phenomena, perhaps explaining why no reservoir deposits have been found. Still, the lack of these deposits could also be explained by recent deepening and widening of the canyon, which would remove the deposits from the canyon's steep slopes.

Certainly some of these lava dams existed for decades or even hundreds or thousands of years since the evidence points to at least five dams that failed catastrophically, perhaps even instantaneously, releasing huge outburst floods of water and rock. Deposits have been recognized along the river corridor with debris containing clasts as large as one hundred feet in diameter and found six hundred fifty feet above the modern river channel. These deposits contain angular basalt clasts that likely were part of the failed dam. Imagine the power and size of a flood that was able to move material of this size so high up onto the canyon walls. Age dating on these outburst flood deposits shows that catastrophic failures occurred between five hundred twenty-five thousand and one hundred thousand years ago.

Younger dam remnants are found inset and juxtaposed against older remnants. This speaks clearly to the enormous cutting power of the Colorado River and its ability to remove obstacles set in its path. Lava dams up to twenty-three hundred feet high and eighty-four miles long are significant obstacles to the river, but the Colorado River rebuffs such

A spectacular lava cascade in a side canyon upstream from Whitmore Wash. Photograph by Wayne Ranney

obstacles and works relentlessly to remove them from its path. A dam likely became compromised when a reservoir filled with water and began pouring over the top. Or perhaps the tremendous hydrostatic pressure that developed within the dam caused the river gravel to be eroded from beneath it, causing the dam to collapse from below. In either scenario, the catastrophic failure of a dam released its decades-long catchment. Another outpouring of lava would start the process all over again, possibly in another locality within this ten-mile stretch of river.

Picture for a moment a time in the not-too-distant past in Grand Canyon. On a cold winter's night, rivers of red-hot lava pour over the canyon rim, cascading into the icy Colorado, filling the canyon with basalt and hot, steamy vapor. Imagine seeing the downstream channel run dry as water became slowly trapped behind a lava dam. Eventually, if the dam was strong and durable, water filled the reservoir behind it and formed a huge lake upstream. Waterfalls would then pour over the top of the various dams—eight hundred feet, twelve hundred feet, even twenty-three hundred feet high—and all set within the colorful walls at the bottom of the canyon. Imagine the view from the rim the moment a lava dam catastrophically failed and a tremendous outburst flood roared through the Grand Canyon, filling it with well over six hundred and fifty feet of

Lava cascades formed natural dams across the path of the Colorado River and likely created reservoirs within the canyon. Some dams may have failed before the reservoir filled, while others likely filled and overflowed, eroding the dams' unstable foundation.

rubble-filled water. The sight and sounds must have been phenomenal. These insights provide a window into the tremendous erosive power of the Colorado River in carving the Grand Canyon.

These musings allow us to learn something else about the canyon-forming process. When a lava dam was formed and then removed, the erosion proceeded only to the pre-dam profile of the canyon wall. This shows that the depth and profile of the canyon are in a kind of equilibrium with the river. When this equilibrium is upset by something akin to a lava dam or a fault-generated knickpoint, the river responds almost instantaneously in an attempt to recover its former profile. Once this profile has been attained, erosion does not appear to progress beyond that preexisting profile. This inquiry may be whispering to us that the canyon may experience discrete periods of active deepening and widening—in response to volcanism, uplift, or climate change—separated by periods when the canyon just sits there and doesn't change very much. I believe we are living in a period when the canyon is just sitting there. But all of this could change with the next uplift event or the next ice age.

An interesting story is beginning to emerge. The canyon may have been significantly deepened within the last 4 million years. Yet, it may undergo certain periods of time when it just sits there silent and still. When someone asks, Is the Grand Canyon getting deeper? perhaps we should appropriately respond, yes and no, or maybe. Not right now, but certainly sometime in the future.

Remnants of lava dams still cling to canyon walls. Photograph by Wayne Ranney

Epilogue

By now we may begin to appreciate the difficulty in trying to decipher the evolution of the Colorado River and its Grand Canyon. It might appear impossible to look through so much time and so much erosion to piece together a coherent story for how the great gorge was carved. I often ask Grand Canyon geologists, "If you could stand at the base of the Grand Wash Cliffs 6 million years ago, what would you see?" To me, the answer to this one question would reveal so much about the Colorado River's history and that of the Grand Canyon. (The only geologist to respond clearly answered, "A thousand-foot waterfall," thus signaling his support for an exclusively young Grand Canyon.)

In the 150-plus years of scientific study at Grand Canyon, a few truths can be offered in providing a satisfactory explanation for its enigmatic origins. A word exercise can be used to help us clarify and understand its complex history. With this exercise we are given a chance to tell the story of Grand Canyon's formation, but the challenge is to tell the greatest "truth" using the fewest words. Our explanation can expand as we continue

The Colorado River flows beneath Toroweap Overlook into Lava Falls in western Grand Canyon. Photograph by Bob and Sue Clemenz

the exercise, but we must still say something that is agreed upon by most geologists without veering off too much into the what-ifs or maybes.

To begin, let's use one sentence to say something about how the Grand Canyon formed. What one sentence tells the most truth about Grand Canyon's origin?

The Colorado River, or some ancestor to it, is responsible for the carving of the Grand Canyon.

This is the simplest, easiest way to begin to explain Grand Canyon's origin because all geologists agree on the close relationship between the history of the river and that of the canyon. We could stop right here, and some people would be completely satisfied to know that this was how the Grand Canyon was formed. But for those who want to know a bit more, what one additional sentence could we add to our simplified explanation that would tell the greatest truth? It might read:

To date, geologists have been unable to determine a precise age for the canyon's beginning and what specific processes worked to create it, although many aspects of its history are now well known.

In our attempt to decipher this story, we now introduce the important idea that there are some unknowns. We have explained that a precise age of the canyon is still elusive and that various possibilities exist that could have shaped it. This simple truth frees us to explore some broader concepts in the canyon's development rather than be tied into any specific details, which in many instances serve only to confuse us. Is there another sentence that can be added at this point that would speak the greatest truth? Let's try:

In fact, the many geologists who have devoted their careers to studying the canyon cannot resolve its age more precisely than somewhere between 70 and 6 million years (although the younger age is the one most often given), and they still debate whether it was formed rather rapidly or over a much more extended period of time.

Now we know just how broad the parameters are in our understanding of the canyon. This sentence serves to say what the extreme limits of the canyon's age may be and that a variety of processes could be involved in its formation. At this point, we can add two sentences to our growing list of truths. This allows us to know something about the larger history of the region that is accepted by nearly all geologists:

All geologists agree that the sea last withdrew from the Grand Canyon region about 80 million years ago, leaving a low-lying, subdued landscape upon which the initial river systems in the region were established. During the ensuing 50 million years, this river system flowed northeast in the opposite direction of the modern Colorado River.

The larger geologic study of the western United States reveals the evidence for this truth. Southwestern Arizona was an area of highlands where rivers drained northeast toward the lowlands that are today located on the Colorado Plateau. Virtually no geologist debates this fact given the overwhelming evidence from many lines of reasoning. An identical geologic setting is found today on the eastern slope of the Andes Mountains, where the headwaters of the Amazon are located. The next sentence adds to these thoughts and is an important qualifier to the puzzle:

It is still unknown whether these northeast-flowing streams have any direct relationship to the present-day position and configuration of all or part of the Colorado River in Grand Canyon, or if portions of the Grand Canyon's depth or extent were formed at this early time.

Moderately deep paleocanyons on the Hualapai Plateau contain deposits that were derived in northeast-flowing streams, but these deposits originated in a different drainage system. We cannot know if these northeast-flowing rivers positioned parts of the modern Colorado River, or if parts of the Grand Canyon were cut at this time. What we do know is that the plateau was elevated at this time, setting the stage for the future deepening of the canyon. This is what follows.

The area that is now called the Colorado Plateau began to be uplifted beginning about 70 million years ago, with additional evidence for two later episodes of uplift.

The uplift history of the plateau and Grand Canyon region is still not fully known, but evidence is found for three periods of uplift for which the relative importance of each is still unknown. Perhaps all three were involved in creating the canyon. A time period with little evidence on the landscape ensues:

About 30 million years ago, the source area for the northeast-flowing streams began to subside through faulting and erosion. The effect this had on the rivers is unknown because very few deposits were laid down or preserved. The rivers here at this time may have become ephemeral, ponded, buried in sand, reversed, or relocated. A deposit in northeastern Arizona suggests that large portions of the Colorado Plateau may have been covered in a vast sand desert, burying or neutralizing these rivers. Rivers here simply disappeared between 30 and 16 million years ago.

The time period between 30 and 16 million years ago reveals the least evidence for what might have occurred in the Grand Canyon area. Little sediment is left to tell us what happened during this time, which may in itself serve as a crude kind of evidence. Perhaps rivers on the plateau were diminished because of an arid climate. Did they later reestablish their courses toward interior basins, or were they reversed such that they found a route to the southwest? We might have a possible answer when we realize:

Interior drainage existed on the northern Colorado Plateau before 30 million years ago and in the adjacent Basin and Range and southern Colorado Plateau after 16 million years ago. This might suggest that interior drainage was present in the Grand Canyon region during this enigmatic time of drainage reversal.

Interior drainage seems to be prevalent throughout much of the history of the Colorado Plateau and might be an indication for the setting that existed across a whole spectrum of time. It could also explain how drainage reversal commenced. We know fairly well what happened next:

Beginning 16 million years ago, the last vestiges of the Mogollon Highlands were destroyed by faulting, and numerous basins developed southwest of the Grand Canyon. The Muddy Creek Formation accumulated west of the plateau edge in an interior basin between 16 and 6 million years ago, but it does not contain any obvious debris from the modern Colorado River. At the same time, the Bidahochi Formation east of the Grand Canyon may have been the site of a lake or numerous ponds. Both of these basins ceased to accumulate sediment after 6 million years ago, and the processes that led to a cessation of deposition are still unresolved.

Two basins existed on either side of the Grand Canyon between 16 and 6 million years ago, and there is some evidence to suggest that some early incarnation of the Grand Canyon may have been positioned between the two. An integration event connected the two basins, and the modern Colorado River was born about 6 million years ago. Three different processes have been proposed as likely candidates for this integration event:

The integration of the Colorado River by headward erosion, basin spillover, or karst collapse is what most likely formed the river we see today. Headward erosion is a favored theory historically, but closed-basin spillover is now documented for the lower Colorado River below Hoover Dam. The specific process that served to rapidly deliver water to the lower river remains speculative, although any of the three processes could have accomplished it.

The modern Colorado River has been a feature on the landscape for only the last 6 to 5 million years, and deposits along the lower river and near the Salton Sea now seem to verify this date. Since the modern river came into existence, Grand Canyon has likely become deeper and wider, explaining why so little evidence remains for its early history and why it looks like such a raw work of nature. The "supersizing" of the canyon is the result of certain processes that act upon it:

Recent studies suggest that the Grand Canyon may have become significantly deeper and wider in the last 4 million years. This was accomplished by the integration of the river, which made a hydrologic connection between the precipitation-rich Rocky Mountains and the low desert basins near the Gulf of California. The upwelling of hot mantle material beneath the crust at Grand Canyon, with consequent uplift and faulting in the western canyon, may have also served to help deepen the canyon.

Lastly, some rather spectacular events have put a final touch on the Grand Canyon, although it would be wrong to think its evolution is over. Volcanoes, lava flows and dams, and outburst floods are documented from the western canyon. Although no humans were around, imagine what it must have been like to see a red-hot lava flow cascading down the walls of Grand Canyon to an ice age river.

One of the last major events to occur in the canyon's long history is the eruption of spectacular basalt lava flows and cinder cones within the canyon and upon its rims. Lava flows cascaded into the Grand Canyon within the last seven hundred thousand years and formed numerous lava dams and perhaps reservoirs behind them. Huge outburst floods have been documented, showing that some of these lava dams failed catastrophically. These events reveal bursts of catastrophic power by the Colorado River to remove newly formed obstacles in its path and to deepen and shape the Grand Canyon.

The Grand Canyon of the Colorado River continues to inspire and enchant legions of scientists and visitors who gaze at it from its spellbinding rims. Photograph by Gary Ladd

Although the development and carving of the Grand Canyon is a complex story that has been very difficult to unveil, we have identified the most pertinent parts of the story. This allows for a relatively brief history of the Colorado River and the Grand Canyon that can be concisely given:

The Colorado River, or some ancestor to it, is responsible for the carving of the Grand Canyon. To date, geologists have been unable to determine a precise age for the canyon's beginning and what specific processes worked to create it, although many aspects of its history are now well known. In fact, the many geologists who have devoted their careers to studying the canyon cannot resolve its age more precisely than somewhere between 70 and 6 million years (although the younger age is the one most often given), and they still debate whether it was formed rather rapidly or over a much more extended period of time.

All geologists agree that the sea last withdrew from the Grand Canyon region about 80 million years ago, leaving a low-lying, subdued landscape upon which the initial river systems in the region were established. During the ensuing 50 million years, this river system flowed northeast in the opposite direction of the modern Colorado River. It is still unknown whether these northeast-flowing streams have any direct relationship to the present-day position and configuration of all or part of the Colorado River in Grand Canyon, or if portions of the Grand Canyon's depth or extent were formed at this early time. The area that is now called the Colorado Plateau began to be uplifted beginning about 70 million years ago, with additional evidence for two later episodes of uplift.

About 30 million years ago, the source area for the northeast-flowing streams began to subside through faulting and erosion. The effect this had on the rivers is unknown because very few deposits were laid down or preserved. The rivers here at this time may have become ephemeral, ponded, buried in sand, reversed, or relocated. A deposit in northeastern Arizona suggests that large portions of the Colorado Plateau may have been covered in a vast sand desert, burying or neutralizing these rivers. Rivers here simply disappeared between 30 and 16 million years ago. Interior drainage existed on the northern Colorado Plateau before 30 million years ago and in the adjacent Basin and Range and southern Colorado Plateau after 16 million years ago. This might suggest that interior drainage was present in the Grand Canyon region during this enigmatic time of drainage reversal.

Beginning 16 million years ago, the last vestiges of the Mogollon Highlands were destroyed by faulting, and numerous basins developed southwest of the Grand Canyon. The Muddy Creek Formation accumulated west of the plateau edge in an interior basin between 16 and 6 million years ago, but it does not contain any obvious debris from the modern Colorado River. At the same time, the Bidahochi Formation east of the Grand Canyon may have been the site of a lake or numerous ponds. Both of these basins ceased to accumulate sediment after 6 million years ago, and the processes that led to a cessation of deposition are still unresolved.

The integration of the Colorado River by headward erosion, basin spillover, or karst collapse is what most likely formed the river we see today. Headward erosion is a favored theory historically, but closed-basin spillover is now documented for the lower Colorado River below Hoover

Dam. The specific process that served to rapidly deliver water to the lower river remains speculative, although any of the three processes could have accomplished it.

Recent studies suggest that the Grand Canyon may have become significantly deeper and wider in the last 4 million years. This was accomplished by the integration of the river, which made a hydrologic connection between the precipitation-rich Rocky Mountains and the low desert basins near the Gulf of California. The upwelling of hot mantle material beneath the crust at Grand Canyon, with consequent uplift and faulting in the western canyon, may have also served to help deepen the canyon.

One of the last major events to occur in the canyon's long history is the eruption of spectacular basalt lava flows and cinder cones within the canyon and upon its rims. Lava flows cascaded into the Grand Canyon within the last seven hundred thousand years and formed numerous lava dams and perhaps reservoirs behind them. Huge outburst floods have been documented, showing that some of these lava dams failed catastrophically. These events reveal bursts of catastrophic power by the Colorado River to remove newly formed obstacles in its path and to deepen and shape the Grand Canyon.

The Vermilion Cliffs near Lees Ferry expose strata believed to have once covered the Grand Canyon area. Photograph by Gary Ladd

Glossary

antecedent stream—One that flowed in its present course prior to the development of the existing topography

anticline—A fold that is upwardly convex such that it forms an arch in rock strata

asthenosphere—That part of the earth's mantle below the rigid lithosphere that behaves like a fluid due to its heat and pressure

basalt—Volcanic rock with about 50 percent silica

basin—The depressed drainage or catchment area of a stream or lake

Basin and Range—A physiographic province west of the Colorado Plateau characterized by a series of parallel mountains and valleys and created about 16 million years ago

bedrock—Any consolidated rock that underlies the rocks or sediment being described

carbonate—A compound of carbon and oxygen, typically found as limestone or dolomite

Cenozoic—The most recent geologic era, from 65 million years ago to the present

clast—An individual fragment or grain in a sedimentary rock formed by the disintegration of some larger rock

Colorado Plateau—A physiographic province centered on the Four Corners states that was gradually uplifted such that the sedimentary rocks remain nearly horizontal

compression—A system of forces or stresses that tend to shorten but thicken rocks

conglomerate—Rounded, water-worn fragments of pebbles cemented together by other minerals

consequent stream—One that follows a course that is a direct consequence of the original slope on which it developed

contact—The place or surface where two different formations come together

cross-bed—Laminations of strata that are oblique to the main planes of stratification

crust—The hard outermost layer of the earth differentiated from the mantle by its composition

depositional environment—The surface conditions such as geographic setting, climate, and transport medium that affect the nature of sedimentary deposits

detrital—Pertaining to grains or clasts

differential uplift—A process whereby one part of the earth's crust rises higher or faster than a part adjacent to it

dip—The angle at which a stratum or any planar feature is inclined from the horizontal

drainage—Any area where water is removed by downslope flow

drainage divide—The topographic border between adjacent drainage basins or watersheds

drainage reversal—The process whereby a river changes its flow direction

ephemeral stream—One that flows intermittently

equilibrium—When the phases of any system do not undergo any change in property with the passage of time

era—A division of geologic time of the highest order

erosion—The group of processes whereby rock is loosened or dissolved and moved from any part of the earth's surface

escarpment—A steep face terminating a high surface; a cliff

extension—The pulling apart of the earth's crust

fault—A fracture along which there has been displacement of the earth's crust

floodplain—The portion of a river valley that is flooded during excessive runoff

fold—A bend in strata

formation—The primary unit of rock layers

granite—An intrusive or plutonic rock rich in silica

headward erosion—Where a stream lengthens its valley either by surface runoff near its headwaters or by subsurface flow in springs (sapping)

Holocene—A geologic epoch consisting of the last ten thousand years

interior basin—A valley, usually with sedimentation, having no outlet to the sea

karst—Any region underlain by limestone where caverns may form and collapse to create sinkholes

knickpoint—Points of abrupt change in the longitudinal profile of a stream

lacustrine—Pertaining to lakes and lake environments

Laramide Orogeny—A mountain-building episode from 70 to 40 million years ago that raised the Rocky Mountains and the Colorado Plateau

lava—Molten rock erupted onto the earth's surface

limestone—Sedimentary rock composed mostly of the mineral calcite

lithology—Pertaining to rock type; in other words, sandstone, shale, and so on

lithosphere—The rigid outer layer of the earth that includes the crust and the upper portion of the mantle

mantle—The layer of the earth's interior below the crust and above the core

marine—Pertaining to the sea or oceanic environment

mature—Landscapes in which maximum development has been reached

Mesozoic—The third era of geologic time, lasting from about 251 to 65 million years ago

Mid-Cenozoic Orogeny—A crustal disturbance located in central Arizona and lasting from about 30 to 20 million years ago

Mogollon Highlands—A mountain range formerly present southwest of Grand Canyon

monocline—A one-limbed flexure in which the beds are flat-lying, except the flexure itself

mudstone—A fine-grained sedimentary rock that includes clay, silt, and sand clasts

Neogene—A time period in the Cenozoic era lasting from 23 million years ago to the present

obsequent stream—One that flows in the opposite direction of the dip of the strata or the tilt of the surface

orogeny—A mountain-building event

outcrop—The exposure of bedrock or strata projecting out from the overlying cover of soil or detritus

Paleogene—A time period in the Cenozoic era lasting from 65 to 23 million years ago

paleogeography—The study of geography through geologic time

Paleozoic—The second era of geologic time, lasting from about 541 to 251 million years ago

plate tectonics—A theory of global-scale dynamics involving the movement of many rigid plates of the earth's crust

Pleistocene—An epoch of the Quaternary period lasting from about 2 million to 10,000 years ago

Quaternary—The most recent geologic period, lasting from about 2 million years ago to the present

relief—The range of topographic elevation within a specific area

resequent stream—A river or stream that began as a consequent stream but was eroded to a lower surface

rift—A deep fracture or break in the earth's crust where two plates pull apart

Rocky Mountains—A physiographic province that was uplifted during the Laramide Orogeny

sandstone—A sedimentary rock containing a large quantity of quartz sand grains

sapping—A process in which groundwater flows out of the ground and causes rocks to become undercut

scarp retreat—Gradual retreat of a cliff face by erosion

shale—A fine-grained sedimentary rock composed of clay and silt grains that splits readily into thin layers

sorting—A descriptive term used to indicate the degree of similarity in a sediment with respect to the size of the grains

spillover—A process whereby a basin fills with water, eventually spilling over its lowest rim

spreading center—The place in the ocean where rifts form in the earth's crust

strata—The layering found within sedimentary rocks regardless of thickness

stream piracy—A process by which the headwaters of a steeper-gradient stream erode headward and capture a lower-gradient stream

subangular—A measure of roundness in gravel in which definite effects of wear are shown; fragments retain their original form and the faces are virtually untouched, but the edges and corners are rounded off to some extent

subduction—Tectonic process in which a dense oceanic plate dives beneath another piece of crust due to plate convergence

subsequent stream—One that has grown longitudinally along belts of soft strata

superposed stream—One that was emplaced on a new surface and maintained its course as it eroded down into preexisting structures

syncline—A fold that is concave upward such that it forms a sag in rock strata

thermochronology—A technique used to understand the thermal evolution of rocks

unconformity—A surface where there is a gap in the rock record

uplift—The elevation of a part of the earth's crust

upwarp—A specific area that has been uplifted

Scientific Bibliography

Babenroth, Donald L., and Arthur N. Strahler. "Geomorphology and Structure of the East Kaibab Monocline, Arizona and Utah." *Geological Society of America Bulletin* 56, no. 9 (1945): 107–150.

Beard, L. S., K. E. Karlstrom, R. A. Young, and G. H. Billingsley, eds. "CRevolution 2—Origin and Evolution of the Colorado River System." Workshop abstract. *U.S. Geological Survey Open-File Report 2011–1210* (2011). Available at http://pubs.usgs.gov/of/2011/1210/.

Blackwelder, Eliot. "Origin of the Colorado River." *Geological Society of America Bulletin* 45, no. 3 (1934): 551–566.

Cooley, Maurice E., and E. S. Davidson. "The Mogollon Highlands—Their Influence on Mesozoic and Cenozoic Erosion and Sedimentation." *Arizona Geological Society Digest* 6 (1963): 7–35.

Crow, Ryan, Karl Karlstrom, Yemane Asmerom, Brandon Schmandt, Victor Polyak, and S. Andrew DuFrane. "Shrinking of the Colorado Plateau via Lithospheric Mantle Erosion: Evidence from Nd and Sr Isotopes and Geochronology of Neogene Basalts." *Geology* 39, no. 1 (2010): 27–30.

Davis, Stephen J., William R. Dickinson, George E. Gehrels, Jon E. Spencer, Timothy F. Lawton, and Alan R. Carroll. "The Paleogene California River: Evidence of Mojave-Uinta Paleodrainage from U-Pb Ages of Detrital Zircons." *Geology* 38, no. 10 (2010): 931–934.

Davis, William Morris. "An Excursion to the Grand Canyon of the Colorado." *Harvard College, Bulletin of the Museum of Comparative Zoology* 38, Geological Series V, no. 4 (1901): 107–201.

Dorsey, Rebecca J., Bernard A. Housen, Susanne U. Janecke, C. Mark Fanning, and Amy L. F. Spears. "Stratigraphic Record of Basin Development within the San Andreas Fault System: Late Cenozoic Fish Creek–Vallecito Basin, Southern California." *Geological Society of America Bulletin* 123, no. 5–6 (2011): 771–793.

Dutton, Clarence E. "Tertiary History of the Grand Cañon District." *U.S. Geological Survey Monograph* 2 (1882).

Elston, Donald P., and Richard A. Young. "Cretaceous-Eocene (Laramide) Landscape Development and Oligocene-Pliocene Drainage Reorganization of Transition Zone and Colorado Plateau, Arizona." *Journal of Geophysical Research* 96, no. B7 (1991). 12389–12406.

Emmons, Samuel F. "The Origin of the Green River." *Science* 6, no. 131 (1897): 20–21.

Fenton, Cassandra R., Robert H. Webb, Philip A. Pearthree, Thure E. Cerling, and Robert J. Poreda. "Displacement Rates on the Toroweap and Hurricane Faults: Implications for Quaternary Downcutting in the Grand Canyon, Arizona." *Geology* 29, no. 11 (2001): 1035–1038.

Flowers, R. M., B. R. Wernicke, and K. A. Farley. "Unroofing, Incision and Uplift History of the Southwestern Colorado Plateau from Apatite (U-Th)/He Thermochronometry." *Geological Society of America Bulletin* 120 (2008): 571–587.

Goldstrand, Patrick M. "Tectonic Development of Upper Cretaceous to Eocene Strata of Southwestern Utah." *Geological Society of America Bulletin* 106 (1994): 145–154.

Gregory, Herbert E. "Geology of the Navajo Country: A Reconnaissance of Parts of Arizona, New Mexico, and Utah." *U.S. Geological Survey Professional Paper* 93 (1917).

Gregory, Herbert E. "Colorado River Drainage Basin." *American Journal of Science* 245, no. 11 (1947): 694–705.

Hamblin, W. K. "Late Cenozoic Lava Dams in the Western Grand Canyon." *Geological Society of America Memoir* 183 (1994).

Hill, Carol A., and Wayne D. Ranney. "A Proposed Laramide Proto-Grand Canyon." *Geomorphology* 102, no. 3–4 (2008): 482–495.

Holm, Richard F. "Cenozoic Paleogeography of the Central Mogollon Rim–Southern Colorado Plateau Region, Arizona, Revealed by Tertiary Gravel Deposits, Oligocene to Pleistocene Lava Flows, and Incised Streams." *Geological Society of America Bulletin* 113 (2001): 1467–1485.

House, P. K., Philip A. Pearthree, and Michael E. Perkins. "Stratigraphic Evidence for the Role of Lake Spillover in the Inception of the Lower Colorado in Southern Nevada and Western Arizona." In *Late Cenozoic Drainage History of the Southwestern Great Basin and Lower Colorado River Region: Geologic and Biotic Perspectives*, edited by M. C. Reheis, R. Hershler, and D. M. Miller. *Geological Society of America Special Paper* 439 (2008): 335–353

Hunt, Charles B. "Cenozoic Geology of the Colorado Plateau." *U.S. Geological Survey Professional Paper* 279 (1956).

Hunt, Charles B. "Geologic History of the Colorado River." *U.S. Geological Survey Professional Paper* 669-C (1969): 59–130.

Ives, Lieutenant Joseph Christmas. *Report upon the Colorado River of the West*. House Executive Document 90. 36th Congress, 1st session. (1861): 93–112.

Johnson, Douglas Wilson. "A Geological Excursion in the Grand Canyon District." *Boston Society of Natural History Proceedings* 34 (1909): 135–161.

Karlstrom, Karl E., Ryan Crow, L. J. Crossey, D. Coblentz, and J. W. Van Wyck. "Model for Tectonically Driven Incision of the Younger Than 6 Ma Grand Canyon." *Geology* 36, no. 11 (2008): 835–838.

Koons, Donaldson. "Geology of the Eastern Hualapai Reservation." *Plateau* 20, no. 4 (1948): 53–60.

Lee, Willis T. "Geology of the Lower Colorado River." *Geological Society of America Bulletin*, 17, (1906): 275-283.

Longwell, Chester R. "How Old Is the Colorado River?" *American Journal of Science* 244, no. 12 (1946): 817–835.

Lovejoy, Earl M. P. "The Muddy Creek Formation at the Colorado River in Grand Wash: The Dilemma of the Immovable Object." *Arizona Geological Society Digest* 12 (1980): 177–192.

Lucchitta, Ivo. "Early History of the Colorado River in the Basin and Range Province." *Geological Society of America Bulletin* 83 (1972): 1933–1948.

Lucchitta, Ivo. "Late Cenozoic Uplift of the Southwestern Colorado Plateau and Adjacent Lower Colorado River Region." *Tectonophysics* 61 (1979): 63–95.

Lucchitta, Ivo. "History of the Grand Canyon and of the Colorado River in Arizona." *Arizona Geological Society Digest* 17 (1989): 701–715.

Lucchitta, Ivo, Richard F. Holm, and Baerbel K. Lucchitta. "A Miocene River in Northern Arizona and Its Implications for the Colorado River and Grand Canyon." *GSA Today* 21, no. 10 (2011): 4–10.

McDougall, Kristin, Richard Z. Poore, and Jonathan C. Matti. "Age and Paleoenvironment of the Imperial Formation Near San Gorgonio Pass, Southern California." *Journal of Foraminiferal Research* 29, no. 1 (1999): 4–25.

McKee, Edwin D., and McKee, Edwin H. "Pliocene Uplift of the Grand Canyon Region—Time of Drainage Adjustment." *Geological Society of America Bulletin* 83 (1972): 1923–1932.

McKee, Edwin D., Richard F. Wilson, William J. Breed, and Carol S. Breed, eds. "Evolution of the Colorado River in Arizona." *Museum of Northern Arizona Bulletin* 44 (1967).

Newberry, John Strong. *Report upon the Colorado River of the West*. 36th Congress, 1st session, House Executive Document No. 90, pt. 3 (1861).

Pederson, Joel. "Colorado Plateau Uplift and Erosion Evaluated Using GIS." *GSA Today* 12, no. 8 (2002): 4–10.

Pederson, Joel, Karl Karlstrom, Warren Sharp, and William McIntosh. "Differential Incision of the Grand Canyon Related to Quaternary Faulting—Constraints from U-Series and Ar/Ar Dating." *Geology* 30, no. 8 (2002): 739–742.

Pederson, J. L., J. C. Schmidt, and M. D. Anders. "Pleistocene and Holocene Geomorphology of Marble and Grand Canyons, Canyon Cutting to Adaptive Management." In *Quaternary Geology of the United States: INQUA 2003 Field Guide Volume*, edited by D. J. Eastbrook. Desert Research Institute, Reno, Nev., (2003): 407–438

Peirce, H. Wesley. "An Oligocene (?) Colorado Plateau Edge in Arizona." *Tectonophysics* 61 (1979): 1–21.

Polyak, Victor, Carol A. Hill, and Yemane Asmerom "Age and Evolution of the Grand Canyon Revealed by U-Pb Dating of Water Table–type Speleothems." *Science* 319 (2008): 1377–1380.

Potochnik, Andre R., and James E. Faulds. "A Tale of Two Rivers: Tertiary Structural Inversion and Drainage Reversal across the Southern Boundary of the Colorado Plateau." Rocky Mountain Section Meeting. *GSA Field Trip Guidebook*, (1998): 149-174.

Powell, John Wesley "Exploration of the Colorado River of the West and Its Tributaries." *Smithsonian Institution Annual Report*, (1875).

Ranney, Wayne D. R. "Geologic History of the House Mountain Area, Yavapai County, Arizona." M.S. thesis, Northern Arizona University, (1988).

Robinson, H. H. "A New Erosion Cycle in the Grand Canyon District, Arizona." *Journal of Geology* 18, no. 8 (1910): 742–763.

Scarborough, Robert B. "Cenozoic Erosion and Sedimentation in Arizona." *Arizona Geological Society Digest* 17 (1989): 515–537.

Schmidt, Karl-Heinz. "The Significance of Scarp Retreat for Cenozoic Landform Evolution on the Colorado Plateau." *Earth Surface Processes and Landforms* 14 (1989): 93–105.

Spencer, Jon E., and P. Jonathan Patchett. "Sr Isotope Evidence for a Lacustrine Origin for the Upper Miocene to Pliocene Bouse Formation, Lower Colorado River Trough, and Implications for Timing of Colorado Plateau Uplift." *Geological Society of America Bulletin* 109 (1997): 767–778.

Strahler, Arthur N. "Geomorphology and Structure of the West Kaibab Fault Zone and Kaibab Plateau, Arizona." *Geological Society of America Bulletin* 59, no. 6 (1948): 513–540.

Walcott, C. D. "Study of a Line of Displacement in the Grand Cañon of the Colorado, in Northern Arizona." *Geological Society of America Bulletin* 1 (1890): 49–64.

Wernicke, Brian. "The California River and Its Role in Carving Grand Canyon." *Geological Society of America Bulletin* 123, no. 7–8 (2011): 1288–1316.

Young, Richard A., and William J. Brennan. "Peach Springs Tuff: Its Bearing on the Structural Evolution of the Colorado Plateau and Development of Cenozoic Drainage in Mohave County, Arizona." *Geological Society of America Bulletin* 85 (1974): 83–90.

Young, Richard A., and Edwin H. McKee. "Early and Middle Cenozoic Drainage and Erosion in West-Central Arizona." *Geological Society of America Bulletin* 89 (1978): 1745–1750.

Young, Richard A., and Earle E. Spamer, eds. "The Colorado River: Origin and Evolution." Grand Canyon Association Monograph no. 12 (2001).

Young, Richard A. "Pre-Colorado River Drainage in Western Grand Canyon: Potential Influence on Miocene Stratigraphy in Grand Wash Trough." In *Late Cenozoic Drainage History of the Southwestern Great Basin and Lower Colorado River Region: Geologic and Biotic Perspectives*, edited by M. C. Reheis, R. Hershler, and D. M. Miller. *Geological Society of America Special Paper* 439 (2008): 319–333.

EARLIER POPULAR BOOKS CONCERNING GRAND CANYON'S ORIGIN

Darton, Nelson Horatio. *Story of the Grand Canyon of Arizona.* Fred Harvey Company, Kansas City, Mo. 1917.

Maxson, John H. *Grand Canyon: Origin and Scenery.* Grand Canyon Natural History Association, Grand Canyon, Az. 1962.

McKee, Edwin D. *Ancient Landscapes of the Grand Canyon Region.* Published by Edwin D. McKee. 1931.

Photograph by Helen Ranney

About the Author

Wayne Ranney is a geologist, educator, and guide who first became interested in earth history while working as a backcountry ranger at the bottom of the Grand Canyon (1975 to 1978). Since that time, his life has revolved around the canyons, rivers, and red rock stratigraphy found in the American Southwest. He received both his bachelor's and master's degrees from Northern Arizona University and enjoys international and domestic travel, backcountry hiking, river rafting, photography, taking long road trips, and watching the sky.

Wayne has taught geology at Yavapai and Coconino Community Colleges and Northern Arizona University, and leads numerous hiking and rafting expeditions throughout the Southwest for the Grand Canyon Field Institute and the Museum of Northern Arizona. He is a popular lecturer locally and on a variety of international expeditions with TCS and Starquest Expeditions. He has visited more than eighty countries and lectured on all seven continents. Wayne's other award-winning books include *Ancient Landscapes of the Colorado Plateau, Sedona Through Time,* and *The Colorado Plateau: A Geologic Perspective*. He and his wife, Helen, live in Flagstaff but are often found exploring the rim or inner depths of their beloved Grand Canyon. You can visit Wayne's website at: www.wayneranney.com and his blog at: www.earthly-musings.blogspot.com/.

Index